NATIONAL ACADEMY PRESS

The National Academy Press was created by the National Academy of Sciences to publish the reports issued by the Academy and by the National Academy of Engineering, the Institute of Medicine, and the National Research Council, all operating under the charter granted to the National Academy of Sciences by the Congress of the United States.

COMPLEX MIXTURES
Methods for In Vivo Toxicity Testing

Committee on Methods for the
In Vivo Toxicity Testing
of Complex Mixtures
Board on Environmental Studies and Toxicology
Commission on Life Sciences
National Research Council

NATIONAL ACADEMY PRESS
Washington, D.C. 1988

NATIONAL ACADEMY PRESS, 2101 Constitution Ave., NW, Washington, DC 20418

NOTICE: The project that is the subject of this report was approved by the Governing Board of the National Research Council, whose members are drawn from the councils of the National Academy of Sciences, the National Academy of Engineering, and the Institute of Medicine. The members of the committee responsible for the report were chosen for their special competences and with regard for appropriate balance.

This report has been reviewed by a group other than the authors according to procedures approved by a Report Review Committee consisting of members of the National Academy of Sciences, the National Academy of Engineering, and the Institute of Medicine.

The National Academy of Sciences is a private, nonprofit, self-perpetuating society of distinguished scholars engaged in scientific and engineering research, dedicated to the furtherance of science and technology and to their use for the general welfare. Upon the authority of the charter granted to it by the Congress in 1863, the Academy has a mandate that requires it to advise the federal government on scientific and technical matters. Dr. Frank Press is president of the National Academy of Sciences.

The National Academy of Engineering was established in 1964, under the charter of the National Academy of Sciences, as a parallel organization of outstanding engineers. It is autonomous in its administration and in the selection of its members, sharing with the National Academy of Sciences the responsibility for advising the federal government. The National Academy of Engineering also sponsors engineering programs aimed at meeting national needs, encourages education and research, and recognizes the superior achievements of engineers. Dr. Robert M. White is president of the National Academy of Engineering.

The Institute of Medicine was established in 1970 by the National Academy of Sciences to secure the services of eminent members of appropriate professions in the examination of policy matters pertaining to the health of the public. The Institute acts under the responsibility given to the National Academy of Sciences by its congressional charter to be an adviser to the federal government and, upon its own initiative, to identify issues of medical care, research, and education. Dr. Samuel O. Thier is president of the Institute of Medicine.

The National Research Council was organized by the National Academy of Sciences in 1916 to associate the broad community of science and technology with the Academy's purposes of furthering knowledge and advising the federal government. Functioning in accordance with general policies determined by the Academy, the Council has become the principal operating agency of both the National Academy of Sciences and the National Academy of Engineering in providing services to the government, the public, and the scientific and engineering communities. The Council is administered jointly by both Academies and the Institute of Medicine. Dr. Frank Press and Dr. Robert M. White are chairman and vice chairman, respectively, of the National Research Council.

This project has been funded by the National Institute of Environmental Health Sciences under Contract NO1-ES-45058 with the National Academy of Sciences. The contents of this document do not necessarily reflect the views and policies of the National Institute of Environmental Health Sciences, and an official endorsement should not be inferred.

Library of Congress Catalog Card Number 87-28178

International Standard Book Number 0-309-03778-6

Printed in the United States of America

Committee on Methods for the In Vivo Toxicity Testing of Complex Mixtures

Subcommittee on Modeling and Biostatistics

RONALD E. WYZGA, Electric Power Research Institute, Palo Alto, California (*Chairman*)
PAUL FEDER, Battelle Columbus Laboratories, Columbus, Ohio
DANIEL KREWSKI, Health and Welfare Canada, Ottawa, Ontario, Canada
STEPHEN LAGAKOS, Harvard University, Boston, Massachusetts
HAROLD N. MACFARLAND, Toxicology Consultant, Victoria, British Columbia, Canada
ELIZABETH H. MARGOSCHES, Environmental Protection Agency, Washington, D.C.
FINBARR O'SULLIVAN, University of California, Berkeley, California
TODD THORSLUND, ICF/Clement Associates, Washington, D.C.
MARVIN SCHNEIDERMAN, National Research Council, Washington, D.C.
JOHN VAN RYZIN, Columbia University, New York, New York
BERNARD WEISS, University of Rochester Medical Center, Rochester, New York

Advisors, Consultants, and Contributors

RICHARD BULL, Washington State University, Pullman, Washington
GAIL CHARNLEY, ICF/Clement Associates, Washington, D.C.
CHRISTOPHER COX, University of Rochester, Rochester, New York
DAVID GROTH, Robert A. Taft Laboratory, Cincinnati, Ohio

National Research Council Staff

RICHARD D. THOMAS, Study Director
BRUCE K. BERNARD, Project Director
EVELYN E. SIMEON, Project Secretary
KATHI COCHRAN-REDMOND, Assistant Project Secretary
ALISON KAMAT, Bibliographer

Technical Contract Monitor, National Institute of Environmental Health Sciences

BERNARD A. SCHWETZ, Systemic Toxicology Branch, Toxicology Research and Testing Program, National Toxicology Program, Research Triangle Park, North Carolina

Board on Environmental Studies and Toxicology

National Research Council Staff

DEVRA LEE DAVIS, Director, Board on Environmental Studies and Toxicology
JACQUELINE PRINCE, Staff Assistant

Preface

The National Institute of Environmental Health Sciences asked the National Research Council to evaluate the toxicity testing of mixtures and make recommendations for improving that testing. The National Research Council responded to the request by appointing 16 scientists to serve on the Committee on Methods for the In Vivo Toxicity Testing of Complex Mixtures, in the Board on Environmental Studies and Toxicology of the National Research Council's Commission on Life Sciences. The committee membership represented the disciplines of toxicology, pathology, biochemistry, analytic chemistry, environmental sciences, biostatistics, and epidemiology.

Toxicity testing is a subject of major current concern to both the scientific and regulatory communities and has been the subject of several recent workshops, reports, and discussions. Consequently, many of the committee members were concerned about our ability to "break new ground" or to provide significant new insights for either the toxicologic methods for testing mixtures or the use of resulting data to protect human and environmental health.

After reviewing both the toxicologic and epidemiologic data on the adverse health effects of mixtures, the committee concluded that a new approach, rather than new methods, is the primary need. This report describes our rationale for this conclusion and presents a series of recommended strategies for testing mixtures.

As the chairman and vice-chairman of this committee, we want to thank Drs. Morton Lippmann, Robert A. Scala, Edo D. Pellizzari, and Ronald Wyzga for chairing task groups that focused on concepts for analyzing human exposure to complex mixtures, testing strategies and methods, sampling and chemical characterization, and interpretation and modeling of toxicity-test results, respectively. Working with this distinguished committee was an honor and a pleasure, and we thank the members for their contributions and hard work in completing this report. Dr. Devra Davis deserves special thanks for the constant encouragement and insight that she communicated to a diverse and large committee. On behalf of the committee, we would also like to thank Drs. Bruce K. Bernard and Richard D. Thomas for their sustained and patient staff effort in support of the committee.

JOHN DOULL, *Chairman*

EULA BINGHAM, *Vice-Chairman*
Committee on Methods for the In Vivo Toxicity Testing of Complex Mixtures

Contents

Executive Summary

People are seldom exposed to single chemicals. Most substances to which people are exposed, whether naturally or artificially produced, are mixtures of chemicals. Mixtures that are of particular concern include chemicals generated in fire, hazardous wastes, pesticides, drinking water, fuels, and fuel combustion products.

This report was prepared in response to a request from the National Institute of Environmental Health Sciences to the Board on Toxicology and Environmental Health Hazards (now the Board on Environmental Studies and Toxicology) of the National Research Council's Commission of Life Sciences to evaluate problems associated with the determination of the toxicity of mixtures. The mandate of the Committee on Methods for the In Vivo Toxicity Testing of Complex Mixtures was to develop strategies and experimental approaches for evaluating the toxicity of mixtures, on the basis of a selective review of the available literature.

The committee reviewed the epidemiologic evidence of effects of exposure to mixtures and found that much of this evidence was derived from exposures to relatively high doses of substances in the workplace. To detect the effects of environmental exposures to low doses, epidemiologic studies will need better methods for documenting relevant exposures and better ways to avoid misclassification of both exposures and outcomes. Epidemiologic investigations generally are able only to confirm past risks. Nonetheless, they can be useful for estimating risks associated with new exposures that are analogous to conditions previously studied. Owing to the various limitations inherent in the available data on human experience, toxicologic studies of mixtures are essential for estimating human risks and can provide strong clues for anticipating, and hence preventing, illness in the human population.

Current toxicologic test methods are applicable primarily to studies of target organ effects, mechanistic hypotheses, and general systemic toxicity. This report focuses on the effects of exposures of mammals, particularly humans, to mixtures. The committee has not considered the larger issues involved in environmental effects or other aspects of general ecology. It recommends that such broad issues be reviewed by another scientific group, and it suggests that more research is warranted on the effects and mechanisms of toxicity in nonmammals, including aquatic species.

Testing complex mixtures present a formidable scientific problem. The key to attacking this problem lies in the analysis and planning of the strategy or experimental approach. The extent and nature of testing should be guided closely by recognition of what is known and what needs to be learned. Selection of any future testing strategy will be linked strongly to questions related to expected exposure, to the toxic end points of interest, and to the likely use and predictive value of the results.

If the question being posed is related to the effects of a mixture, the strategies invoked involve toxicity testing of the mixture itself. The first step should be a careful consideration of the origin of the mixture, because this will provide some basic information on the complexity and the physical and chemical characteristics of the mixture, which might imply the types of anticipated effects. Care needs to be taken in the sampling process: samples must reflect as closely as possible what humans are potentially exposed to; they must be suitable to the types of assay chosen; and their integrity must be maintained at all points, as one means of ensuring relevance to the human situation. Questions regarding stability of mixtures must be considered, because different components age or decompose at different rates, and that creates uncertainty about what is actually being tested, compared to what humans are exposed to. Established and emerging toxicologic methods are as adequate for evaluating complex mixtures as for evaluating single agents.

Toxic-effects strategies are limited by the lack of animal or cellular models for some effects (e.g., cardiovascular and respiratory disease). Therefore, the committee encourages the continued development of animal models that permit critical studies of alien disease. The search for exposure-related effects in human populations must continue, and the resulting information must be integrated into the development of animal models in toxicology.

One of the central questions associated with complex mixtures is related to finding the primary causative agents of specific health effects. The strategies for identifying causative agents depend heavily on the integration of toxicology with chemical characterization. In light of the complexity and diversity of many mixtures, the complete chemical characterization of every sample under study is neither prudent nor possible.

Bioassay-directed fractionation, which is not used for single agents, is the most useful strategy for studying mixtures. Bioassays of fractions derived from mixtures are used to identify which fractions need chemical characterization; the most toxic fractions are quickly identified.

In dealing with the ability to predict adverse effects of mixtures, the committee distinguished between factors related to effects and factors useful in modeling. The predictive value of an effects strategy centers on the ability to use data from tests of one mixture to predict the likely effects of exposure to a new but similar mixture. Among the effects strategies, comparative-potency studies and matrix testing were designed for such prediction. (Matrix testing involves the identification and manipulation of several variables that are arrayed in a two-dimensional matrix, such as boiling range and aromaticity.) The predictive value of a model strategy is related to the toxicity of components of simple mixtures whose components are all known. In such a case, if the toxicity of each component is known or can readily be determined, it might not always be necessary to test the whole mixture. (This model-driven prediction is described in Chapter 3.)

Simple mathematical models have been used to assess the toxicity of simple mixtures, particularly binary mixtures. Those models suggest that interactions that might be observed at a high dose, such as an experimental dose, do not necessarily occur at lower doses; hence, models might play an important role in permitting the extrapolation of toxicity to lower doses. Although a concentrated effort has been made to define the role of models in assessing the toxicity of more complex mixtures, it is clear that further work is needed to test the validity of models against experimental and epidemiologic data.

Consistency of analytic results between mixtures implies an increase in the predictability of their effects. The more similar two mixtures are chemically, the more similar their toxicologic properties are expected to be. However, even mixtures that are relatively well characterized sometimes have unexpected toxic effects. One must be prepared to look for unexpected results of exposure to complex mixtures, because of the potential disjunction between chemical analysis and biologic effect. Complex chemical mixtures are more likely to produce unexpected results than are individual chemical substances, for several reasons. Mixtures are composed of various substances, exposure to which can be expected to be associated with different toxicities. The constituents of a mixture sometimes combine chemically to produce new compounds with different toxicities. The presence of some materials might mask, dilute, or increase the toxicity of other materials. Such phenomena, referred to as interactions, can amplify or reduce anticipated effects. Moreover, different doses of separate materials might increase the bioavailability of materials that are otherwise nontoxic at the doses present in the mixture.

On the basis of theoretical considerations and its examination of some epidemiologic studies, the committee noted that effects of exposures to agents with low response rates usually appear to be additive. The only examples of interaction that were considered greater than additive occurred in humans exposed to agents, such as cigarette smoke, that alone produced a high incidence of effects. Current quantitative models used to assess cancer risks support these results. However, the committee did not thoroughly review the toxicologic data on additivity assumptions.

The committee recognizes that several important related issues are not discussed in its report. Its discussion of the testing of complex mixtures deals largely with strategies, rather than with detailed testing methods.

1

Introduction

Contaminated drinking water, leachates and volatile substances from hazardous waste sites, and individual diet and life style all involve exposure to thousands of chemicals that can affect the likelihood or course of disease. This report of the Committee on Methods for the In Vivo Toxicity Testing of Complex Mixtures reviews the methodologic requirements for testing the toxicity of these chemical mixtures to humans. It does not review testing of environmental toxicity.

Toxicity testing to predict effects on humans has traditionally studied chemical compounds one at a time, for various reasons: dealing with agents singly has been more convenient to investigators; physicochemical properties of single agents were often readily defined; dosage could be easily controlled; biologic fate could usually be monitored in a straightforward manner; concentrations in air, water, and tissue could be accurately measured; target-organ toxicity was predictable on the basis of experience with agents related to the one in question; and relevant data were often available from human occupational exposures. However, most occupational and environmental chemical exposures are to mixtures, not single agents. Almost all commercial products contain impurities. Some products—such as fuels, pesticides, and coal tar—have dozens of constituents. Tobacco smoke consists of a few thousand chemicals.

Human exposure to complex mixtures has been extensive, and some adverse effects are well documented, especially in terms of chronic disease associated with long exposures. But the extent of human exposure to even well-defined mixtures that are responsible for producing excess disease is generally not known in quantitative terms; and the relative roles of mixture components associated with the effects are rarely known. Furthermore, the relationships between exposure and dose to the target organ are often not well defined in either humans or experimental animals. The quantitative predictive power of toxicity-test data from animal studies for human health effects is often limited by a lack of understanding of interspecies differences in biologically effective dose. Nonetheless, toxicity data on animals provide a key component of primary preventive medicine, in that they provide a rational basis on which to estimate, and hence prevent or reduce, human harm.

For the study of mixtures, testing strategies must consider similarities and differences between mixtures and must use methods that maximize the ability to extrapolate results between mixtures. Toxicologic evaluation of complex mixtures is complicated by the potential for component interactions. Interactions could influence physicochemical properties, dose to a target tissue, biotransformation and elimination, and toxicity. In the absence of adequate human data, data on the toxicity of a complex mixture (or its components) must be acquired with animal (surrogate) models. Use of these models is likely to be complicated by unpredictability of toxicity resulting from interactions, difficulties in extrapolating from high-dose to low-dose exposures, and the ability of several materials to cause the same toxic effect.

Questions asked about complex mixtures might be quite similar to those asked about single entities. Testing strategies designed to answer those questions, however, might be very different. For example, asking whether a specific component of a mixture causes cancer is different from asking whether a single chemical causes cancer. In the first case, a fractionation protocol might be needed, perhaps with short-term in vitro assays to identify the toxic fractions; in the second case, fractionation is not applicable. Environmental mixtures are often bound to a matrix, such as soil, and extraction procedures must be developed to isolate a mixture or its components. However, health data based on separation of a mixture from its matrix might not be relevant to actual human exposures.

Consideration of human experience with complex mixtures is important, because many toxic effects of chemical mixtures were not, and in some cases could not have been, predicted on the basis of experiments with animal models. For example, the extent to which cigarette-smoking and alcohol consumption affect the risk of oral cancer would not have been predicted on the basis of animal experiments, because experimental models for detecting oral cancer due to either cigarette-smoking or alcohol consumption are inadequate. In some cases, further interpretations of available data or additional use of standard tests can provide a basis for anticipating effects. But these cases generally would involve chronic tests that are seldom performed, and the conventional protocols and rodent models are inadequate for eliciting many of the responses seen in epidemiologic studies.

Consideration of chemical-mixture toxicity must also take into account the potential for interaction between components of a mixture. Interaction between components can occur in several ways. Chemicals can interact with each other within a mixture to form one or more new compounds. On exposure of an organism, some components might affect the absorption, distribution, metabolism, or elimination of other components. Such interactions can lead to toxicity greater or less than would be expected on the basis of the individual component toxicities. Interactions that lead to unpredictably greater toxicity might pose public-health problems, particularly in the case of acute exposures to high doses. Most biologically significant interactions observed at the high doses used for chronic toxicity testing in the laboratory, however, are not detectable at the low doses to which the human population is exposed environmentally.

DEFINITIONS

Several terms used throughout this report in the context of interaction are difficult to define, because of their varied interpretations in different disciplines. "Interaction" is a general term that has been applied to toxicity-test results that deviate from dose-additive behavior expected on the basis of dose-response curves obtained with individual agents.

Biologists tend to think of "synergism" as referring to any linear result that is greater than would be expected from addition of doses. Suppose that two chemicals are carcinogenic in the same organ by the same mechanism, for example, by forming adducts at critical sites on DNA. Exposure to both chemicals will result in a tumor rate that reflects simple addition of doses. If, instead, one acts by an indirect mechanism, such as that of a tumor promoter, exposure to both agents might result in a tumor rate that is greater than that expected on the basis of simple addition of doses. Biologists refer to this phenomenon as "synergism." Statisticians describe the situation as multiplicative if the results could be predicted on the basis of multiplication of doses and as synergistic if the results are greater than would be predicted on the basis of the underlying dose-response model. Antagonism is a situation in which the response is less than would be predicted on the basis of simple addition of doses. Such classifications are model-dependent; the credibility of a dose-response model depends on the extent to which is is based on some understanding of the biologic mechanisms underlying the interaction.

"Exposure" refers to the contact of a substance with an object of interest. If the object of interest is a living organism, a distinction must be made between the exposure (e.g., ambient concentration) and the biologically effective dose (commonly known simply as "dose"). The latter is the quantity of material that interacts with the biologic receptor responsible for a particular effect. Because determination of a biologically effective dose requires intimate knowledge of the mechanisms of action, in many instances imperfect knowledge leads to the use of surrogates for this quantity. The less knowledge we have concerning the receptor and the steps leading to the substance-receptor interaction, the less precise are our surrogates (i.e., estimates) for the biologically effective dose and the less predictive they are for estimating doses in other species (e.g., humans).

Current statistical models for describing and extrapolating toxicity data generally were developed for single substances, and their application to mixtures can lead to incorrect inferences when mixture components interact. Several of the most commonly used dose-response models suggest that the risk associated with exposure to a mixture at a "low" dose (i.e., with a response rate of about 0.1–1%) is equal to the sum of the risks associated with exposure to the mixture components separately. However, these results are not only dose-limited, but also model-dependent, and the credibility of the result is closely linked to the credibility of the model. Sound scientific judgment is the key ingredient in the application of any particular model to a specific situation.

STRUCTURE OF THE REPORT

Chapter 2 of this report discusses the availability of various types of qualitative and quantitative data that form the bases of currently used analyses of human exposures to complex mixtures (e.g., epidemiology). It presents several situations in which standard toxicity-testing protocols did not predict adverse health effects of such exposures. Concern about the nature of these situations has led investigators and regulators to question whether the use of standard, largely acute, single-chemical toxicity-testing methods at high exposures is appropriate for mixtures, especially when many of the documented effects of mixtures are attributable to chronic, small exposures.

Chapter 3 restates the fundamentals of toxicity testing of single chemicals in the context of chemical mixtures. Testing of mixtures differs only in extent, not in kind, from testing of single agents. The two critical elements in evaluating complex mixtures are the careful definition of experimental objectives (including the uses to which the resulting data will be put) and the necessity of differentiating exposure from the biologically effective dose. The chapter examines testing strategies to determine optimal approaches to various experimental objectives.

Chapter 4 provides guidelines for the design of protocols for sampling and analyzing chemical mixtures. Analytic strategies should include consideration of the nature of exposure and bioavailability and should supplement toxicity testing. A proposed classification scheme considers degree of complexity, sample origin, physical state, and chemical properties. Chemical characterization of a complex mixture might not be necessary before testing, but predicting effects of chronic exposure and extrapolating test results to other mixtures are often improved by such characterization.

Chapter 5 explores the role that biostatistics and quantitative modeling can play in describing the toxicity of mixtures. This role includes both the development of testing strategies and experimental designs and the interpretation of results, including analysis of interaction. The importance of using mathematical models that are consistent with postulated biologic mechanisms of action is emphasized. The assumptions underlying successfully predictive models can in themselves stimulate additional biologic research into the mechanisms of toxicity. The conclusions derived in the chapter are expanded in appendixes that describe their mathematical bases.

2

Concepts for Analyzing Human Exposure to Complex Chemical Mixtures

People are seldom, if ever, exposed to single chemicals. Instead, most natural and artificially produced substances are mixtures of chemicals, such as those in acid deposition, fire products, hazardous wastes, pesticides, surface drinking water, and the products of fuel distillation or combustion. Appendix A discusses the origins of major categories of the complex mixtures that people encounter today. A few studies of human populations exposed to particular mixtures have demonstrated toxic effects (see Appendix B). Analyses of experimental studies, mostly toxicologic studies in animals, have later verified the toxic nature of a mixture or even identified its specific toxic constituents. However, the number of mixtures identified through this process of epidemiologic hypothesis generation followed by toxicologic verification is small. This chapter discusses many of the problems inherent in the process, not the least of which is that such a system is premised on confirming past human hazards, rather than seeking to predict, and hence prevent, such hazards.

The impetus, therefore, for studying complex mixtures is the need to generate reliable predictions on which to base estimates of potential health risks for humans and ultimately to guide exposure controls. Except for clinical trials of therapeutic agents, deliberate high-dose exposure of large populations to materials of unknown potential for harm is unacceptable in modern society. The prediction of human health risks uses a "weight-of-evidence" approach that considers data from various sources—epidemiology, analyses of accidental or intentional human exposures, animal tests, cell culture studies, metabolic and mutagenic assays, and so on—and weights the data from all those sources in relation to their relevance to human populations (WHO, 1981; Ballantyne, 1985).

For any chemical exposure, an effect of concern must be identified and a dose-response relationship must be estimated. The dimensions of the dose-response relationship usually prove particularly troublesome, primarily because estimates of dose (biologically effective dose, as described in Chapter 1) are poor in most collections of epidemiologic evidence. Even the quantification of exposure can be difficult.

This chapter begins with a discussion of factors that affect dose and dose surrogates (using the case of inhalation exposure to demonstrate the complexity of the process), including biologic, analytic, and statistical problems that typically arise in dose characterization. With that discussion as a base, the second and third sections discuss the qualitative and quantitative problems associated with attempting to characterize human exposures. Those discussions include not only the complexity of the materials to which people are exposed and the complexity of the exposed populations, but also the interactive effects that can ensue from exposure to mixtures of biologically active substances.

In theory, the most relevant effects data are those generated from human exposures. In practice, however, such data are usually unavailable or so poorly characterized (by confounding exposures, inadequate dose quantification, etc.) that data from experimental animal studies must be relied on. The fourth section of this chapter discusses various types of data that are commonly collected on both human and animal exposures, in an attempt to show how they might be compared, and suggests reasons for potential divergence.

The final section of this chapter presents eight examples of human experience with complex mixtures that involve a number of failures to anticipate human harm and includes discussion of types of data that could have been helpful in predicting problems. (For readers interested in more complete accounts, Appendix B contains more detailed chronologies of those examples.)

GENERAL CONCEPTS

Exposure represents a contact between a chemical agent and an object. This report is concerned with exposure of living organisms, particularly humans. Exposure, which may be considered an independent variable, is different from the dose that reaches a target organ or tissue (the biologically effective dose or, simply, dose), which may be considered a dependent variable. The magnitude of a dose is a reflection of the magnitude of an exposure modified by a series of intervening processes, including inhalation or ingestion; transfer of inhaled or ingested material across epithelial membranes of the skin, respiratory tract, and gastrointestinal tract; transport via circulating fluids to target tissues; and uptake by the target tissues. Those processes in turn vary with a subject's activity level, age, sex, health status, and inherent constitutional makeup with respect to race or species, size, and so on.

When people are subjected to chronic or repetitive exposures to chemical agents, other important factors affect the magnitude of the dose. If the agent or its effect on the target tissue is cumulative and clearance of the agent or recovery from its effect is slow, the dose is a function of the total cumulative uptake. But if the agent is rapidly eliminated or its effect is rapidly and completely reversible on removal from exposure, the dose is a function of the rate of uptake. (See the 1985 NRC report *Epidemiology and Air Pollution* for other discussion of these concepts.)

Dose and Dose Surrogates

We can seldom measure dose directly. The two most commonly used surrogates for dose are exposure and biologic fluids from exposed persons, which can provide evidence that absorption has taken place or that biologically effective doses have been delivered. Dose surrogates can include the presence of the chemical of interest itself in blood, urine, or exhaled air or the presence of metabolites or enzymes induced by uptake of the chemical. Therefore, assessing exposure and its resulting dose is a complex process involving a set of multidisciplinary activities (NRC, 1985).

The generation of data useful for dose estimation through sampling is a function of two processes. The first is the design of sampling protocols to obtain biologically relevant data; the second is the actual collection of samples themselves. Techniques for sampling air, water, food, and biologic fluids are relatively well developed, as are techniques for sample separation from copollutants, media, and interferences and for quantitative analysis. The design of appropriate sampling protocols requires an understanding of the critical variables that determine the magnitude of dose. We need to know when, where, how long, and at what rate and frequency to collect samples to obtain data relevant to the exposures of interest; obtaining such data requires knowledge of the temporal and spatial variabilities of exposures and concentrations and knowledge of metabolic pathways and kinetics. We seldom have enough information of these kinds to guide our collection of samples.

The use of exposure as a dose surrogate introduces the further problem of inadequate quantitative data, particularly in the case of retrospective studies. Assignments to the various exposure groups, either "exposed" versus "nonexposed" or graded exposures, are usually based on work histories, questionnaires, and some knowledge of process conditions that justify judgments about relative magnitudes of exposure. All these methods of assessing exposure are indirect and lead to a degree of misclassification (NRC, 1986a). However, when they are used in conjunction with measurements of biologic fluids or of exposure, the rate of misclassification can be estimated.

The issues just described are related to quantifying dose on the basis of exposure. These issues are more complicated when the exposure is to complex mixtures, rather than single entities.

INDICATORS OF EXPOSURE TO COMPLEX MIXTURES AS DOSE SURROGATES

Because it is impractical or impossible to measure the concentrations of all the constituents of a complex environmental mixture in an exposure environment or in body fluids, exposures or retained amounts are sometimes estimated from data on one or several constituents or their metabolites. The use of indicators can be rationalized on the basis either that the retention of the constituent is representative of the retention of one constituent of interest or that the toxicity of the marker is representative of the toxicity of the mixture.

An example of a representative constituent is benzo[a]pyrene (BaP), which is one of the most frequently used indicator compounds for complex organic mixtures. However, BaP alone does not always accurately reflect the carcinogenic potency of a complex mixture. Its concentration in air has been correlated with relative risks of lung cancer for coal-tar workers, coke-oven workers, and others (Speizer, 1983). It is a potent animal carcinogen, and it can be routinely analyzed by analytic techniques of proven reliability (IARC, 1983). Although BaP is likely to be present to some degree in most environmental mixtures, its mass or biologic activity can be highly variable (NRC, 1985).

An epidemiologic study of roofers exposed to very high concentrations of BaP found some excess lung cancer among those exposed for more than 20 years (Hammond et al., 1976). The standardized mortality ratios (SMRs) for those exposed for 9–19, 20–29, 30–39, and 40 or more years were 92, 152, 150, and 247, respectively. (The smoking histories of the workers were not known.) Airborne BaP concentrations measured in roofing operations range from 14 $\mu g/m^3$ in the roof-tarring area to 6,000 $\mu g/m^3$ in the coal-tar roofing kettle area (Sawicki, 1967). The amounts of BaP recovered from masks, which were worn by a minority of roofers studied, indicated an average of 16.7 μg of BaP inhaled per day (Hammond et al., 1976). Another group exposed to high concentrations of airborne BaP were British gasworkers, who inhaled about 30 μg/day (Lawther et al., 1965).

The mainstream smoke of a cigarette contained about 3.5×10^{-2} μg of BaP in 1960 (Kotin and Falk, 1960). Thus, a 2-pack/day smoker inhaled about 1.4 μg/day. Ambient urban air in 1958 contained about 6 ng/m^3, which could account for an inhaled mass of about 120 ng/day (i.e., 0.12 μg/day). Ambient rural air has about 10% as much BaP (Faoro, 1975).

The research cited above on roofers exposed to BaP indicates that their ambient exposures are considerably higher than those of cigarette smokers. However, the rate of lung cancer is much higher for cigarette smokers than for roofers. This indicates that BaP at best is a crude indicator of the carcinogenic potential of a complex mixture. This example is pertinent because it suggests that additional compounds in a complex mixture can be important in the expression of the response and that the physical phase of the indicator compound can play a role.

There are also indicators for pollutant mixtures with other health effects. The U.S. Environmental Protection Agency (U.S. EPA, 1986) uses ozone (O_3) as an indicator for the presence of photochemical oxidants, nitrogen dioxide (NO_2) as an indicator for the various nitrogen oxides, sulfur dioxide (SO_2) as an indicator for the various sulfur oxides, and the mass concentration of suspended particulate matter (PM) for all pollutants present as droplets or solid particles. In the committee's judgment, the validity of these indicator chemicals as hazard indexes varies from very good (O_3) to very questionable (PM).

Effects of Matrix and Physical State on Dose

Risk of health effects is generally considered to be a function of both toxicity and exposure, so it is important to consider bioavailability to determine the extent of potential exposure once a potentially toxic agent has been identified by chemical analysis. Bioavailability defines the biologically effective dose, a variable fraction of the administered dose, and should be considered in estimating the body burden of a chemical. Bioavailability can be affected (either increased or decreased) by agents or conditions that alter release, uptake, metabolism, distribution, or biologic effects. With knowledge of bioavailability, one can better understand the biologic end points and characterize the toxicity of a complex mixture with regard to additivity, antagonism, and synergism. If bioavailability assays are conducted immediately after chemical characterization, the risk assessor will have important end points with which to develop a hazard evaluation by minimizing the uncertainty and clarifying exposure information (U.S. EPA, 1985).

An example of how bioavailability can differ is the case of 2,3,7,8-tetrachlorodibenzo-*p*-dioxin (TCDD), whose bioavailability from environmental samples can vary from less than 0.1% up to 85%, depending on the matrix to which it is bound, the media from which it entered the environment, the duration of the binding to environmental substrates, and the presence of other compounds in the mixture. This can be best illustrated by studies conducted with TCDD-contaminated soil samples from Times Beach, Missouri, and Newark, New Jersey (De Caprio et al., 1986; McConnell et al., 1984; Umbreit et al., 1986b; Van der Berg et al., 1985). Both soils were contaminated by a number of agents in addition to TCDD. The biologic assay of the contaminated soils involved administering TCDD at equivalent concentrations to rats and guinea pigs and monitoring for appearance of TCDD-related symptoms. A comparison revealed that both soils induced cytochrome P-450 enzyme systems, but only the Times Beach soil induced the TCDD syndrome and death in guinea pigs. These differences in response were associated with the ease of extraction of TCDD and other agents (shaking in solvent followed by column chromatography) from the Times Beach samples, in contrast with the rigorous methods (48–72 hours of exhaustive Soxhlet extraction) required to extract similar chemicals from Newark soils (Umbreit et al., 1986b). The bioavailability of components was the decisive factor in the evaluation of the toxicologic hazard of these environmental samples. An underlying concern of researchers and regulators is whether the risk assessor should consider only the presence of xenobiotics in a medium and the toxicity of these xenobiotics (assuming exposure) or should attempt to use bioavailability data to complete the exposure assessment.

The organic content appeared to be higher and binding sites appeared to be more abundant in Newark soil than in Times Beach soil. Whether it can be generalized that matrices with higher organic content have greater binding affinity than similar matrices with lower organic content is not clear. Poiger and Schlatter (1980) and Rappe et al. (1986) showed that fly ash, sediments, and carbonaceous materials bind several organic compounds, but have a high affinity for chlorinated dibenzodioxins and dibenzofurans. The presence of solvents or the continued release of solvents at a site might aid in the percolation of compounds through the soil and increase binding to soil particles. This phenomenon has been hypothesized as an explanation for some of the soil binding of polychlorinated biphenyls (PCBs) in Japan.

Some of the constituents of a mixture might alter the absorption of others, or the mixture might alter gastrointestinal transit time and thus affect absorption of several compounds, including nutrients (Hollander, 1981). Analytic methods are available to determine differential uptake; indeed, Bandiera et al. (1984) have recently demonstrated selective hepatic and adipose retention of specific chlorinated dibenzofurans from complex mixtures of them and PCBs found at Yusho, Japan.

Bioavailability varies with route of exposure. The oral, dermal, and respiratory routes differ in selectivity of uptake, rejection, and storage. The bioavailability of a constituent of a complex mixture will depend on that constituent's solubility, volatility, charge, and concentration and on those of other compounds in the mixture. Research efforts and resources should be aimed at differential bioavailability of constituents of complex mixtures in different physical states. The case of fuels and fuel exhausts or effluents might be slightly simpler, because the compounds (those not particle-bound) are in a nonpolar liquid medium (highly lipid-soluble) or in a vapor phase; bioavailability is a function of membrane effects and differential uptake in the absence of a matrix (Klaasen, 1986).

Bioavailability also depends on the physical state of the mixture. Several investigators have studied the bioavailability of dioxins and PCBs from liquid or semisolid media and have found reasonable agreement between theoretical and actual biologic values. However, in the case of mixtures bound to solid substrates, there are marked differences from site to site (Umbreit et al., 1986a).

FACTORS INFLUENCING RELATIONSHIPS BETWEEN EXPOSURE AND DOSE

Estimating human pulmonary damage involves making a number of assumptions about effective dose. The problems of estimating the effective dose of inhaled materials are described here to illustrate the complexity of the dose estimation process.

The surface and systemic uptake of chemicals from inhaled air depends both on the physical and chemical properties of the chemicals and on the anatomy and pattern of respiration within the airways. Gases and vapors rapidly contact airway surfaces by molecular diffusion. For water-soluble gases, dissolution or reaction with surface fluids on the airways facilitates removal from the airstream. Highly water-soluble vapors, such as SO_2, are almost completely removed in the airways of the head, and very little penetrates into lung airways. For water-insoluble compounds, surface uptake is limited, and thus the decline in concentration with depth in the lung can be low. For such compounds, the greatest uptake is often deep in the lung, where the residence time and surface areas are the greatest.

For airborne particles, a critical characteristic affecting surface deposition patterns and efficiencies is particle size. A model to predict the percentage deposition of particles in various regions of the respiratory tract was developed by the Task Group on Lung Dynamics (ICRP, 1966) of the International Commission on Radiological Protection.

Almost all the mass of airborne particulate matter is found in particles with diameters greater than 0.1 μm. The penetration of airborne particles into the lung airways is determined primarily by convective flow—that is, the motion of the air in which the particles are suspended—because these particles are small in relation to the airways in which they are suspended. Some deposition by diffusion does occur for particles smaller than 0.5 μm in small airways, whereas for particles larger than 0.5 μm, deposition by sedimentation occurs in small to midsized airways. For particles with aerodynamic diameters greater than 2 μm, particle inertia is sufficient to cause particle motion to deviate from the flow streamlines; that results in deposition by impaction on surfaces downstream of changes in flow direction, primarily in midsized to large airways that have the highest flow velocities. The extent of deposition on limited surface areas within the large airways is of special interest with respect to dosimetry and the pathogenesis of chronic lung diseases, such as bronchial cancer and bronchitis.

Quantitative aspects of particle deposition are summarized in a recent air quality criteria document (U.S. EPA, 1982). The deposition efficiencies in different regions of the human respiratory tract are highly variable among healthy subjects. Additional variability results from structural changes in the airways associated with disease processes. Generally, these changes involve airway narrowing or localized constrictions that act to increase deposition and concentrate it on limited surface areas.

The preceding discussion of particle deposition is based on the assumption that each particle has a specific size. However, hygroscopic particles enlarge considerably as they take up water vapor in the airways (NRC, 1978).

Materials that dissolve in the mucus of the conductive airways or the surfactant layer of the alveolar region can rapidly diffuse into the underlying epithelia and the circulating blood, thereby gaining access to tissues throughout the body. Chemical reactions and metabolic processes can occur within the lung fluids and cells, limiting access of the inhaled material to the bloodstream and creating reaction products with either greater or less solubility and biologic activity. Few generalizations about absorption rates are possible for individual chemicals, let alone for complex chemical mixtures.

Particles that do not dissolve at deposition sites can be translocated to remote retention sites by passive and active clearance processes. Passive transport depends on movement on or in surface fluids lining the airways. There is a continuous proximal flow of surfactant to and onto the mucociliary escalator, which begins at the terminal bronchioles, where it mixes with secretions from Clara and goblet cells. Within midsized and large airways, there are additional secretions from goblet cells and mucous glands, producing a thicker mucous layer having a serous subphase and a more viscous gel layer. The gel layer, lying above the tips of the synchronously beating cilia, is found in discrete plaques in smaller airways and becomes more nearly continuous in the larger airways. The mucus reaching the larynx and the particles carried by the mucus are swallowed and enter the gastrointestinal tract.

The total transit time for particles deposited on ciliated conductive airways extending to the terminal bronchioles varies from about 2 to 24 hours in healthy nonsmoking humans. Macrophage-mediated particle clearance via the bronchial tree takes place over a period of several weeks. The particles deposited in alveolar zone airways are ingested by alveolar macrophages within about 6 hours; but turnover of the macrophages normally takes several weeks. At the end of several weeks, the particles not cleared to the bronchial tree via macrophages have been incorporated into epithelial and interstitial cells, from which they are slowly cleared by dissolution or as particles passing through pleural and eventually hilar and tracheal lymph nodes via lymphatic drainage pathways. Clearance rates for these later phases depend strongly on the chemical nature of the particles and their sizes, with half-times ranging from about 30 days to 1,000 days or more.

All the characteristic clearance times cited above refer to inert, nontoxic particles in healthy lungs. The presence of toxicants can drastically alter clearance times. Inhaled materials that affect mucociliary clearance rates include cigarette smoke (Albert et al., 1974, 1975), sulfuric acid (Lippmann et al., 1982; Schlesinger et al., 1983), O_3 (Phalen et al., 1980), SO_2 (Wolff et al., 1977), and formaldehyde (Morgan et al., 1984). Macrophage-mediated alveolar clearance is affected by SO_2 (Ferin and Leach, 1973), NO_2 (Schlesinger, 1986), sulfuric acid (Schlesinger, 1986), O_3 (Phalen et al., 1980; Schlesinger, 1986), and silica dust (Jammet et al., 1970). Cigarette smoke is known to affect the later phases of alveolar zone clearance in a dose-dependent manner (Bohning et al., 1982). Both clearance pathways and rates can be affected by these toxicants, which thus influence the distribution of retained particles and their dosimetry.

Given the sparseness of current knowledge and dose-related clearance rate data, pulmonary retention of complex chemical mixtures cannot be predicted. That might pose greater difficulties for interpreting exposures to complex mixtures in occupational settings, where the exposures could be high enough to alter clearance pathways and rates. For general population exposures at lower concentrations, it might be reasonable to assume, as a first approximation, that pulmonary clearance rates are normal.

The above discussion illustrates the importance of determining characteristics of biologic systems directly affected by exposure. The physiology of normal and abnormal or diseased organs should be well characterized.

EFFECT OF POPULATION AND EXPOSURE COMPLEXITY ON QUALITATIVE ASSESSMENT

Although the choice of appropriate indicators for environmental measurements of exposure and the nature of bioavailability are complicated issues, perfect knowledge of these would still leave many other problems in assessing exposure of human populations unresolved. Epidemiologic studies necessarily assess the effects of both exposures and other factors that occur in the complex and variable social and physical environment in which people live and work. Few exposures, even to single chemicals, occur in the absence of other environmental factors.

Complexity of Exposure

In occupational or general environmental settings, it is important to know the temporal and concentration characteristics of exposures. Short-term high-level exposures might occur in some industrial settings, for example, during recharge or cleaning of normally sealed batch reaction chambers. Long-term low-level exposures occur in the general environment; for example, a broad spectrum of the population, some of whom might be in compromised health, receive relatively low-level exposures to air pollutants continuously. Exposures to mixtures of chlorination byproducts in drinking water are usually at low levels, but occur frequently (NRC, in press). Because exposure can start in infancy and continue throughout a lifetime, the cumulative dose can be substantial.

Similar types of exposure can occur among different groups of exposed persons at different levels and frequencies. For example, x irradiation for diagnostic purposes occurs at relatively low levels and infrequently in the general population and usually results in a cumulative lifetime dose of under 10 rads (Radford, 1986). But x rays are also used frequently at much higher levels to control tumor growth in cancer patients. In the past, x irradiation was also used to treat many other medical conditions that were not life-threatening, such as severe acne, tinea capitis, and ankylosing spondylitis. As with x rays, exposure to industrial and environmental chemicals can vary in intensity and frequency across the population.

Important considerations in designing epidemiologic and toxicologic studies are the origin and degree of inherent complexity of mixtures, as well as the ways in which exposure to them can be modified or controlled to decrease the risk of disease. Adverse effects of exposure to proprietary blends of well-characterized substances often can be ameliorated simply by removing or substituting for toxic constituents, but this strategy is obviously ineffective for mixtures whose toxic factors are integrally related to their intended use. Products of combustion from a variety of sources and chlorination byproducts resulting from the disinfection of drinking water are examples of this type of mixture (NRC, 1986a,b, in press). Identifying all the toxic components in these cases is usually not possible. Even if the toxic components could be identified, removing individual constituents is seldom technologically feasible (two notable exceptions are reduction of the nicotine content of tobacco and the caffeine content of coffee and cola drinks); control consists of reducing exposure to all constituents, toxic and nontoxic alike. Cigarette smokers, for example, who switch to filtered brands reduce exposure to the total mixture, but not necessarily to the key toxic ingredients; the obvious way to decrease exposure to the toxic ingredients is to decrease or cease smoking. Decreases in the exposure of workers and the general public to coke-oven emissions are accomplished by controlling all the emissions, not only the most toxic fractions. Concentrations of organic substances in drinking water can be controlled by activated-carbon filtration. Activated carbon removes the relatively small fraction of chemicals that are toxic, as well as the bulk of the ones that are not. Thus, in designing toxicologic testing approaches, one must consider whether the toxic components are intrinsic to the mixture and difficult to control as individual substances or whether they are easily identifiable and potentially separable from the mixture.

Complexity of Populations

In populations, the exposure setting and nature of the affected groups can profoundly influence the type and severity of an adverse effect. Are exposures limited primarily to the occupational setting, or is the general population at risk? If the latter, special attention is warranted for high-risk groups, such as infants, the elderly, pregnant and lactating women, the poor, and others whose sensitivity might be increased by previous exposure, chronic medical conditions, or inherited characteristics. Even if groups are not identified as being at especially high risk, age, sex, and normal variation of metabolic characteristics can have important consequences for the reaction of exposed persons.

Nutritional status and preexisting medical conditions in the host organism might have a profound influence on the response to environmental exposures. For example, a variety of micronutrients appear to confer protection against chemical carcinogens (NRC, 1983a). That can occur through direct interaction of the micronutrients with a potential carcinogen or indirectly through enzymatically mediated detoxification, whereby the micronutrient is a critical enzyme cofactor. Selenium, for example, is a necessary cofactor of glutathione peroxidase, which serves as a reducing agent for potentially carcinogenic peroxides (Ip and Sinha, 1981). Some micronutrients that might confer protection from carcinogens are vitamin A, many of the retinoids, carotene, vitamin C, and indoles (NRC, 1983a). It is important, then, to consider nutritional status of test animals in evaluating complex mixtures.

Some human groups might be especially sensitive to particular environmental exposures. People might be sensitized to the dermal effects of irritating chemicals by extensive previous exposures. Asthmatics could suffer severe bronchoconstriction on exposure to sulfur oxides at concentrations that would have little or no impact on nonasthmatics.

Interactive Effects of Multiple Exposures

Toxicologic testing of complex mixtures can answer some questions raised by observations among humans. As discussed below, experimental designs must carefully consider exposure conditions, host characteristics, appropriate health end points, and potential interactions between constituents of mixtures and multiple types of exposure.

Interactive effects of concurrent exposure to two or more known risk factors have been evaluated in a number of epidemiologic studies. Some examples are lung cancer resulting from exposure to cigarettes and radon daughters (Little et al., 1965; Rajewsky and Stahlhofen, 1966; Pershagen et al., 1987), lung cancer due to cigarette-smoking and asbestos exposures (Selikoff et al., 1968; Berry et al., 1972; Hammond and Selikoff, 1973), and oral cancer from alcoholic-beverage consumption and smoking (McCoy et al., 1980). Statistical modeling of the available epidemiologic data in most cases has not resolved and is unlikely to resolve whether these exposures interact in an additive or a synergistic manner. Even less clear is the most appropriate biologic or physiologic model that can explain the action of multiple environmental insults. Toxicologic testing of joint effects of two or more chemical exposures, under carefully controlled conditions, would help answer many of the outstanding questions.

Toxicologic evaluation of complex mixtures can address many of the questions raised specifically by epidemiologic studies of workplace and other exposures to mixtures. Large-scale analyses of extensive data bases with occupational and mortality data have identified groups of workers in several industries at high risk of cancer and other diseases. Some of these surveys have used mortality records kept by vital-statistics bureaus in several states that now include the usual data on occupation and industry, in addition to underlying cause of death and other items abstracted from death certificates (Milham, 1983, 1985). Several surveys of this type have been conducted in the United States and Britain (see, for instance, Office of Population Censuses and Surveys, 1978; Logan, 1982). Typically in these studies, the proportion of deaths by cause has been analyzed within each of many occupations or industries to indicate the causes of death that might be linked to specific job classifications. The exposure in these studies is limited to inferences based on the occupation or industry entered on the death certificate. Many occupational groups that show excess risk of dying of particular causes have been identified by these surveys and warrant more detailed evaluation with both epidemiologic and toxicologic approaches. (Further discussion of the use and limitations of these surveys can be found in Kazantzis and McDonald, 1986, and Hernberg, 1986.)

For many occupational groups at high risk, specific exposure information is often not available or is too complex to interpret. That is true for several industries with occupations at high risk of cancer that are the subject of several recent monographs from the International Agency for Research on Cancer (IARC) in the series *IARC Monographs on the Evaluation of the Carcinogenic Risk of Chemicals to Humans*. The first 24 volumes of the IARC monograph series restricted attention to the carcinogenic properties of individual chemicals. But starting in 1981, with Volume 25 (*Wood, Leather, and Some Associated Industries*), and continuing with Volume 28 (*The Rubber Industry*, 1982) and Volume 34, Part 3 (*Polynuclear Aromatic Compounds: Industrial Exposures in Aluminium Production, Coal Gasification, Coke Production, and Iron and Steel Founding*, 1984), the monographs have reviewed the carcinogenic risk of employment in selected industrial occupations. The changed focus in this series of reports reflects a growing recognition that exposure to complex mixtures or multiple insults in some occupational settings is not amenable to simple analysis or interpretation on the level of individual chemicals. In these industries, workers who held specific jobs with exposure to complex mixtures are shown to be at increased risk of cancer, although specific exposures cannot be identified. Further chemical characterization of mixtures, followed by their toxicologic testing, would help to clarify excess cancer risk among workers in a number of workplace settings described in these monographs.

The rubber industry (*IARC Monographs,* Volume 28, 1982) provides an excellent illustration of the potential value of further chemical characterizations and toxicologic testing. Through a collective-bargaining agreement, the United Rubber Workers and several U.S. rubber and tire companies funded an extensive series of occupational studies on the risk of cancer and other diseases. These studies revealed excess risk of several cancers within specific job categories. For example, they revealed excess mortality due to lymphocytic leukemia among employees whose jobs might have exposed them to benzene and other solvents. But the epidemiologic data were unable to discriminate further among the solvents. Toxicologic testing of solvent mixtures thus could help to pinpoint the sources of excess risk in this industry and could serve as a basis for removing or replacing offending substances.

Complexity of Health Effects

Cancer is but one end point that should be considered in evaluating health effects of complex mixtures. Similar types of exposure can induce other effects, depending on the conditions of exposure, exposure setting, and host characteristics. Some examples that might be appropriate for toxicologic evaluation include mutagenesis, teratogenesis, neurotoxicity, bronchoconstriction, hepatotoxicity, renal insufficiency, and a variety of dermatologic effects. Systematic review of the epidemiologic literature can help to focus toxicologic testing of particular mixtures on questions directly applicable to the human experience.

CONSIDERATIONS IN QUANTITATION OF HUMAN EXPOSURES IN EPIDEMIOLOGIC STUDIES

Documented effects of environmental chemicals on humans seldom contain quantitative exposure data and only occasionally include more than crude exposure rankings based on known contact with or proximity to the materials believed to have caused the effects. Interpretation of the available human experience requires some appreciation of the uses and limitations of the data used to estimate the exposure side of the exposure-response relation. The discussion that follows is an attempt to provide some relevant background for interpreting the available, retrospective data and for specifying the kinds of data that might be collected in prospective studies.

There are both direct and indirect sources of exposure data and data that can be used to rank exposed subjects into exposure-intensity groups. Among the quantitative direct approaches are direct measurements of external exposures and measurements of exposure via analysis of biologic fluids. The indirect measures generally rely on work or residential histories and data on exposure intensity at each exposure site or on enumeration of the frequency of process upsets or effluent discharges that result in high-intensity short-term exposures.

Ambient Exposure

The problems of exposure characterization differ with the nature of the exposure. Many epidemiologic studies of complex mixtures are based on occupational exposures. These involve repetitive daily exposures lasting up to about 8 hours, with uptake predominantly by inhalation, but sometimes also by skin absorption. Most studies of the health effects of community air pollution involve complex mixtures with highly variable temporal exposure patterns (NRC, 1985). The biologically relevant exposure times can also be highly variable. The temporal patterns of ambient concentration can depend on photochemical reaction sequences and strong gradients between indoor and outdoor concentrations. Ingestion exposures from food and water depend strongly on methods of food preparation, dietary preferences and variety, and other nutritional factors or deficiencies.

Concentrations in Air, Water, and Food

Historical data are occasionally available on the concentrations of materials of interest in environmental media. But the data might or might not relate to the exposures of interest. We must ask some important questions before attempting to use such data:

- How accurate and reliable were the sampling and analytic techniques used in the data collection? Were they subjected to any quality assurance protocols? Were standardized or reliable techniques used?
- When and where were the samples collected, and how did they relate to exposures at other sites? Air concentrations measured at fixed (area) sites in industry might be much lower than those occurring in the breathing zone of workers close to the contaminant sources. Air concentrations at fixed (generally high) community air sampling sites can be either much higher or much lower than those at street level and indoors, owing to gradients in source, strengths, and rates of pollutant decay in indoor and outdoor air.
- What is known or assumed about the ingestion of food or water containing the measured concentrations of the contaminants of interest? Time at home and dietary patterns are highly variable among at-risk populations.

Biologic Sampling Data

Many of the questions that apply to interpretation of environmental media concentration data also apply to biologic samples, especially quality assurance. The time of sampling is especially critical in relation to the times of the exposures and to the metabolic rates and pathways. In most cases, it is difficult to separate the contributions to circulating concentrations from recent exposures and those from long-term reservoirs.

Exposure Histories

Exposure histories themselves are generally unavailable. However, work histories and residential histories can yield information on exposure, as can reliable data on norms with respect to hygiene programs in specific industries. Job histories, as discussed above, are often available in company or union records and can be converted into relative rankings of exposure groups with the aid of long-term employees and managers familiar with the work processes and the history of process changes. Knowledge about materials handled and about tasks performed is critical. So is knowledge about the installation and effectiveness of engineering modification for exposure control. Routine, steady-state exposures might be the most important and dominant exposures of interest in many cases. But for some health effects, occasional or intermittent peak exposures could be of greater or primary importance. In assessing or accumulating exposure histories or estimates, it is important to collect evidence on the frequency and magnitude of the occasional or intermittent releases associated with process upsets.

Some Statistical Issues in Quantitation of Exposure

The preceding discussions have addressed some complications in the assessment of exposure for complex mixtures, including the parameters of exposure-response relations and measurement of exposure. The section below addresses the role of statistical methodology in these issues and highlights problems for which additional methodologic research is desirable.

A number of descriptive and mathematical models have been developed to permit estimation of exposure from knowledge of exposure and one or more of the following factors: translocation, metabolism, and effects at the site of toxic action.

The use of these models for airborne particulate matter generally requires a knowledge of the concentration within specific particle-size intervals or of the particle-size distribution of the chemicals of interest. The most widely used of these models was developed by the International Commission on Radiological

Protection and Measurement (ICRP, 1966). The model describes, in probabilistic terms, the sites of deposition of particles of various sizes in the lungs.

A key analysis issue is the consequence of mismodeling exposure history in estimating or testing an exposure-response relation or association. For example, in a study of the association between one or more air pollutants and the risk of chronic respiratory impairment, we must determine whether cumulative, peak, recent, or some other measure of exposure history is used. In many if not most situations, biologic knowledge is not sufficiently detailed to indicate which measure is most appropriate. We know from general principles that using an incorrect measure can result in biased estimates of the exposure-response relation and in reduced sensitivity (i.e., statistical power) in the detection of an exposure-response association (see, for instance, Inskip et al., 1987).

Another analysis issue is the consequence of inexact measurements of exposure. In most applications, the measured value of exposure is only an approximation of the actual—that is, absorbed—exposure. Two key issues are the effects of inexact exposure on the estimation of the exposure-response relation and the loss of statistical power for detecting an association between exposure and response.

Exposure assessment issues also have important implications for the design of exposure-response studies. Although it often is clear that obtaining exact exposure information would be prohibitively expensive, relatively little is known about how to balance the added value of more precise exposure data with the corresponding increase in cost. This problem can be regarded as one of experimental design, and statistical literature on methods of determining designs that are "optimal" for some prespecified criteria, which account for both cost and accuracy, could play an important role in the design of future exposure-response studies.

PREDICTIVE VALUE OF LABORATORY STUDIES FOR HUMAN EFFECTS OF ENVIRONMENTAL EXPOSURES

The rationale for using animal studies to evaluate the human risks associated with environmental pollutants is that evolution has endowed mammalian species with similar genetic, biochemical, and physiologic makeups. These similarities extend to toxification and detoxification mechanisms and to target sites for the adverse effects of pollutants.

Toxicity testing is based on principles discovered in the study of comparative toxicology, which is founded on the basic disciplines of comparative biology, comparative physiology, and comparative biochemistry. Those disciplines have demonstrated that successful extrapolation of animal toxicity studies to humans requires knowledge not only of general similarities among species, but also of differences between species.

Interspecies comparisons of toxic effects of pollutants can be carried out at a number of levels of varied complexity. These levels include gross end points (such as lethality, tumorigenesis, and mutation induction in bacteria), more descriptive toxicokinetic events (such as absorption, distribution, metabolism, and elimination), and changes at the cellular level. Although molecular toxicology holds great promise, preliminary findings from this emerging field indicate that basic biochemical pathways are very similar for all forms of life and nearly identical for related species. That implies that molecular mechanisms of toxicity are qualitatively the same for related species and that interspecies differences in toxicity are due primarily to quantitative (toxicokinetic) differences in biochemical pathways.

At present, the most productive approach to toxicity testing continues to be bioassays for gross toxic end points in species related to humans (mammals). Agents that have adverse effects in test animals are assumed to be dangerous to humans, although the degree and specific manifestation of toxicity in humans might differ from those in test animals, because of underlying toxicokinetic variables. For that reason, successful extrapolation of animal toxicity data to humans requires comparing animal and human toxicity data and elucidation of toxicokinetic events that are relevant to interspecies variability in toxicity. The process is aided by knowledge gained from surveys of occupationally exposed persons and from both unintentional and carefully controlled experimental human exposures. In rare cases, such as when therapeutic agents are given to critically ill persons, animal and human toxicity data can be compared directly.

Exposure as a Variable

Acute responses to short-term exposures in animal experiments are often difficult to compare with analogous events in human populations. Short-term human exposures to toxic amounts of substances usually result from accidental spills or workplace accidents that often are difficult to characterize. Imagine the difficulty in accurately determining the human LD_{50} for methylisocyanate from the Bhopal tragedy. In addition, most toxicologic research is oriented toward establishing safe concentrations for long-term exposures, which are generally much lower than concentrations that produce acute toxicity.

Human and animal acute responses to short-term exposures have been compared directly for some anticancer drugs (Freireich et al., 1966; for a review, see Calabrese, 1983). In these studies, maximal tolerable doses (MTDs) of anticancer drugs that physicians gave to patients as a normal part of cancer chemotherapy were compared with MTDs for the same drugs in a number of different species. When the doses were expressed per unit area of body surface, all species, including humans, had comparable MTDs.

Most of the agents tested in the above studies were electrophilic alkylating agents or antimetabolites for which there is apparently little interspecies difference in critical toxicokinetic variables. However, many chemicals exhibit substantial interspecies toxicokinetic variability that accounts for differences in toxic effects between humans and test animals. (This topic is discussed in more detail later in this chapter.)

In contrast with the short-term testing mentioned above, most toxicity testing is oriented toward protecting human populations from chronic disease that results from long-term exposures. The most widely studied chronic disease associated with long-term exposures to environmental pollutants is cancer. It is therefore not surprising that studies of carcinogenesis provide the best example of the utility of animal toxicity testing to predict human diseases that result from exposure to environmental pollutants.

The concentration and chronicity of the exposure can influence responses to single agents or to mixtures of agents. Nitrogen dioxide in concentrations present in polluted ambient air (0.05–1.0 ppm) can in the presence of sunlight oxidize organic compounds and produce peroxides and peroxyacetyl nitrate, a strong conjunctival irritant. High concentrations of NO_2, such as those in cigarette smoke (50–100 ppm), can produce ciliastasis and inflammatory reaction in the respiratory tract. When cigarette smoke is inhaled chronically in amounts generated from one to two packs per day for 20 or more years, chronic lesions develop in several organs, including the lung and cardiovascular system, and neoplasms develop that are not predictable from any of the acute lesions. Emphysema and chronic bronchitis develop as a result of chronic prolonged cigarette-smoke inhalation; however, their development in humans cannot be predicted from the acute pulmonary effects associated with cigarette-smoke inhalation in animals (U.S. Surgeon General, 1984). Coronary arterial disease and hypertension—both chronic lesions of the cardiovascular system—cannot be predicted by laboratory studies of cigarette smoke, even with chronic and prolonged exposures (U.S. Surgeon General, 1983). Neoplasms of the lung, larynx, mouth, and esophagus likewise are not produced in animals by inhalation exposure to cigarette smoke (U.S. Surgeon General, 1982).

Complexity of the Agent as a Variable

In general, the simpler the agent, the more likely that the laboratory data will demonstrate changes that are useful as a guide to its actions in humans. The acute effects of NO_2 are an example. NO_2 at concentrations over 40 ppm for 12 hours will produce profound, and even fatal, pulmonary edema in humans, as well as in experimental animals (NRC, 1977, 1981). As the agent becomes increasingly complex, so too does the effect on the animal. Beagles exposed to SO_2 and diluted auto exhaust showed increased airway resistance and decreased lung volume and compliance (Hyde et al., 1978; Bloch et al., 1972), and they developed mild emphysema.

Cigarette smoke is a mixture of great complexity with medical manifestations (NRC, 1986a). Cigarette smoke contains over 3,800 compounds, each in a different concentration (Dube and Green, 1982). It has a particulate phase and a gaseous phase. The nature and amounts of particles generated depend on the volume of the inhaled puff, the temperature of the burning tobacco, its dilution with air, the nature of tobacco, and the distance the smoke must travel from ignited tobacco to target organ. New polycyclic aromatic hydrocarbons are generated by pyrolysis; in addition, new compounds are formed by oxidation, and free-radical formation initiates peroxide formation in unsaturated fatty acid and alkoxyl intermediates. The biologic effects of this galaxy of compounds are virtually impossible to predict from a knowledge of the individual constituents, not only because of their huge number, but also because transient chemicals generated during the processes of pyrolysis, oxidation, and free-radical formation might dissipate or change with time and temperature.

Limitations of Animal Models

In several ways, laboratory animals respond in a manner different from that of humans when exposed to some environmental agents. Allergy and hypersensitivity, common in humans, appear not to affect most laboratory animals in a similar manner. Atmospheric pollutants can irritate and aggravate respiratory abnormalities in people with allergic diatheses or bronchospastic airways. Hypersensitivities to ingested agents—such as aspirin, antibiotics, and other drugs—can commonly produce organic diseases, including pulmonary fibrosis and drug-induced hepatitis, that are not commonly observed in animals. Furthermore, laboratory animals do not commonly develop chronic disease states—such as arteriosclerosis, emphysema, and malignant neoplasms—in response to small exposures to environmental toxicants. For instance, neither emphysema nor arteriosclerosis has been adequately created experimentally as a result of cigarette-smoke inhalation or other agents.

Finally, animal studies are not reflective of human disease when the agent or mixture tested causes a recurrence or exacerbation of an existing disease. People with allergic asthma, chronic nonspecific reactive bronchitis, or acute infections might respond vigorously to inhaled irritant mixtures, whereas healthy animals exposed to similar mixtures will not.

In spite of the previously described difficulties associated with the identification of human carcinogens, 30 chemicals or chemical mixtures and 9 industrial processes have been identified as definitive human carcinogens—studies on humans have confirmed that the agents cause cancer in those who are exposed (IARC, 1982; Vainio et al., 1985; Rall et al., 1987). A review of IARC monographs found 288 chemicals, industrial processes, and complex mixtures for which some data on human carcinogenicity or sufficient evidence of carcinogenicity in experimental animals existed (Vainio et al., 1985). For the purpose of classifying carcinogenic risk in humans, IARC concludes that, in the absence of epidemiologic studies, sufficient evidence of carcinogenicity in animals is reasonable grounds for regarding a chemical as carcinogenic in humans (IARC, 1986). In a study of the adequacy of toxicologic testing of chemicals in commerce, a National Research Council committee reported that most such chemicals are not adequately tested for potential toxicity (NRC, 1983b). Furthermore, many substances cannot be evaluated for carcinogenicity in humans, because adequate data on human exposure are not available (Karstadt and Bobal, 1982). Several chemicals have been shown to be carcinogenic in humans, but cannot be adequately tested in a laboratory setting. A 1975 report (NRC, 1975) reviewed much of the information and stated:

> It may be noted that the organs affected in man are not always the same as those that are found to develop tumors in laboratory animals, nor is the the organ specificity the same in different species of rodents. Nevertheless, the evidence, based on a limited number of carcinogens, suggests that most agents that pose a carcinogenic threat to man will be carcinogenic in laboratory tests on animals. However, this leaves open the possibility that such tests may also identify chemicals carcinogenic to rodents that do not pose such a threat to man.

That report went on to describe a number of examples in which the organ specificity, a qualitative effect, is the same in humans and animals but the magnitude of the effect is decidedly different (aflatoxin B_1 and diethylstilbestrol) or in which the effect in the most sensitive animal species is qualitatively different from that in humans (vinyl chloride) or in any other animal species (benzidine). Of course, these comparisons have been made in the absence of information on molecular dosimetry. However, when study results are compared on the basis of milligrams of substance per kilogram of body weight, both qualitative and quantitative differences between humans and animals are observed (NRC, 1975).

In some circumstances—because of methodologic shortcomings, species differences, or particular end points of interest—animal studies might not be predictive of human carcinogenicity. In studies of SO_2 and suspended particulate matter, animal studies have not adequately reflected the effects of polluted atmospheric air in humans (see Appendix B for details).

The use of laboratory animals has been of limited value in the study of combined exposures, such as exposure to radon daughters and cigarette smoke and to asbestos and cigarette smoke (see Appendix B for details). The effects of radon daughters and of asbestos alone are adequately reflected in animal studies. But in experimental studies of combined exposures in which cigarette smoke is one of the agents, it is not reliably delivered to the target organ in adequate quantities. Adequate investigation of the effects of inhalation exposure to cigarette smoke requires long-term chronic studies, which are expensive and difficult. An effective alternative has been to apply the concentrated extract of tobacco tars to a surrogate organ during exposure to the other agent, such as radon daughters or asbestos. However, the predictive validity of this altered method for the human conditions can be questioned. This strategy could be applied to studies of noncancer effects of other complex mixtures to provide cost efficiency and time-saving. Studies in laboratory animals provide the best means available to anticipate and hence prevent human harm.

Toxicokinetics

Differences in toxicokinetics are often the explanation for interspecies variability in susceptibility to toxicants. Much of the research in this subject consists of studies of efficacy and side effects of drugs. In many cases, molecular events have been elucidated that are critical control points for the toxicokinetics of chemicals. One of the control points is the activity of toxification and detoxification processes. For example, the duration of activity of hexobarbital varies greatly from species to species and is directly correlated with species ability to metabolize the parent compound to an inactive form (Weiss, 1978). Variations in intestinal and dermal absorption can also influence the activity of an agent; sheep and cattle absorb smaller percentages of ingested lead than humans and thus have greater tolerances for dietary lead (Scharding and Oehme, 1973).

After absorption, toxicants often must cross other biologic barriers to reach their targets of action. Species differ in the ability of toxicants to penetrate barriers, such as the placenta (Koppanyi and Avery, 1966) and the blood-brain barrier (Way, 1967; Rall, 1965, 1971). The differences often explain interspecies variability in sensitivities to teratogens and neurotoxins. Species susceptibilities to toxic effects can vary with differences in plasma-protein binding, biliary excretion, and enterohepatic circulation (Albert, 1979; Smith, 1973), all of which can affect the metabolism, distribution, and elimination of toxicants. All these differences have been extensively reviewed by Calabrese (1983, 1986).

SUMMARIES OF EXAMPLES OF HUMAN EXPOSURES TO COMPLEX CHEMICAL MIXTURES

The final section of this chapter contains eight summaries of situations in which there has been an apparent nonconcurrence between human and animal health responses. (For readers interested in additional information, more detailed descriptions are provided in Appendix B.) In the situations discussed, human responses often have differed substantially from what might have been expected on the basis of data from controlled laboratory exposures to pure materials. Furthermore, a review of the abstracts shows that for many cases more careful interpretation of laboratory research presaged these findings of toxic effects in humans. For other cases summarized here, epidemiologic evidence first signaled toxicity, and the search for an animal model continues. Thus, experimental studies can play an important role in predicting human effects and reducing disease.

The detailed case studies (Appendix B) discuss the experimental data that suggest or have helped to establish causal relationships. They also provide information that might have helped, if it had been used or interpreted differently (a judgment admittedly based on hindsight). The purpose of the latter is not to criticize past decisions, but to identify the types of experimental data that are most useful in extrapolation to human disease potential.

Sulfur Dioxide and Suspended Particulate Matter

The sulfur-oxide/particulate-matter complex common to urban areas has been clearly associated with excess daily mortality and respiratory tract morbidity (NRC, 1985; D. V. Bates, personal communication). Only in recent years have animal models for relevant functional and morphometric effects proved useful in developing an understanding of the human health effects, and these have involved using animals and exposure protocols not used in conventional toxicity testing (beagles and dilute auto-exhaust mixtures, guinea pigs and SO_2-particle mixtures passed through a heated furnace, etc.). Other informative studies involved nonconventional functional assays (e.g., particle clearance function in rabbits undergoing brief daily exposures to sulfuric acid aerosols). Collectively, the results of the animal studies and of some studies involving short-term exposures of human volunteers point to one specific component of the mixture—acidic aerosol—as the most likely causal factor for the various demonstrated human health effects.

Lead and Nutritional Factors—Effects on Blood Pressure

Recently, lead (Pb) was identified as an independent causal factor for hypertension in men among a host of dietary cofactors that were previously known or strongly suspected to affect hypertension. The evidence for the independent role of lead comes primarily from NHANES II, a national population-based study that included a substudy on blood lead, blood pressure, and dietary factors. Supporting evidence for the influence of blood lead on blood pressure comes from a series of recent epidemiologic studies of occupationally exposed populations. In retrospect, published data on blood-lead/blood-pressure associations in rats could have been used to predict the human association. One reason that the relation was not recognized sooner through toxicologic or epidemiologic studies was its unusual shape: a plateau of elevation in blood pressure at a relatively low concentration of blood lead. Most toxicologic data are generated at relatively large exposures, and effects that occur only at smaller exposures are likely to be overlooked.

Radon Daughters and Cigarette Smoke

Several investigators have speculated that cigarette-smoking and radon daughters have joint actions in the incidence of lung cancer among miners exposed occupationally. Both radon daughters and cigarette-smoking are known to increase lung-cancer incidence in humans independently, and their joint action produces no more than an additive cumulative incidence. Smokers, however, have a shorter latent period for lung cancer than nonsmokers. One study found that nonsmokers chronically exposed to cigarette smoke have an increased risk of lung cancer if radon is also found in the home (Pershagen et al., 1987). Animal models have been of only limited value in studies of joint action of these two agents, at least for the inhalation model of exposure. Animals in groups large enough for cancer studies cannot be exposed to cigarette smoke in a manner analogous to the way humans are exposed. Topical application of cigarette-smoke condensate to rodent skin provides a means of studying some effects of joint action, but it is of small value as a model for the human effects of concern, because the bioavailability is likely to be different between smoke and smoke condensate.

Asbestos Exposure and Cigarette-Smoking

Asbestos is known to cause asbestosis, mesothelioma, and lung cancer in humans. Human studies have also demonstrated that cigarette-smoking has no apparent effect on the incidence of asbestosis or mesothelioma among those exposed to asbestos (NRC, 1984). But cigarette-smoking does markedly increase the incidence of lung cancer among asbestos-exposed humans, and the increase appears to be multiplicative. There is no suitable animal model for cigarette-smoke-related health effects. Without such a model, animal studies on joint action are not feasible.

Cigarette-Smoking and Alcoholic-Beverage Consumption as Risk Factors in the Etiology of Oral Cancer

There is evidence of joint action of cigarette-smoking and alcohol ingestion in the incidence of oral cancer in humans. The evidence provides a rare example of a demonstrated health effect with available quantification of exposure to both agents. Therefore, it allows a more detailed mathematical analysis of joint action than can be performed in most studies on the effects of mixtures. Even here, however, it is not possible to establish definitively whether the effects of the two agents are additive or multiplicative. The absence of a relevant animal model also limits the ability to address the critical issues in controlled experiments.

Trihalomethanes and Other Byproducts of Chlorination in Drinking Water

This case concerns the epidemiologic basis for the presumption of human risk of cancers of the gastrointestinal and urinary tracts associated with exposure to chlorination byproducts in disinfected drinking water. Of the situations presented here, evidence that exposure to chlorination byproducts increases the risk of cancer in humans is the most difficult to establish. There are several reasons. Increases in human cancer risk are likely to be small, in a relative sense, and are therefore difficult to detect in an epidemiologic setting with many possible confounding exposures. Mixtures of volatile and nonvolatile chlorination byproducts vary geographically and temporally and thus defy unambiguous definition (NRC, in press). The chlorinated organics in the nonvolatile fraction thought to be of greatest toxicologic importance occur in extremely low concentrations and are difficult to detect and measure. Toxicologic evaluations have relied largely on in vitro tests in nonmammalian systems and therefore are of only limited value in suggesting the cancer sites of greatest concern and the magnitude of expected risk. Further evaluation of risk associated with exposure to this complex mixture is a challenge to both epidemiologists and toxicologists. The issue deserves much further attention, because exposures are so widespread and potential absolute risks so high.

Coke-Oven Emissions

Epidemiologic studies have shown an increased incidence of lung and genitourinary tract cancers in connection with exposure to coke-oven emissions, whereas experimental inhalation studies in animals have not produced comparable results. However, technical limitations have prevented exposures of experimental animals to realistic inhalation atmospheres. As a consequence, toxicologic insights have been obtained from studies involving surrogate exposures and cancers in other tissues—in particular, studies in which extracts of particulate emissions were shown to produce skin cancer after topical applications. The animal studies provide opportunities for determining relative potencies among the numerous constituents of the coal-tar complex. Despite the substantial differences in the human and animal data, the effects of coal tar in animals are related to the effects of coke-oven emissions in humans. Furthermore, the data generated by toxicity testing of complex mixtures in animals can be a reasonable predictor of human disease, if the mixture tested in the laboratory is representative of mixtures in the human environment, if studies are properly designed to detect diseases, and if we are aware that the specific diseases and target organs found in test animals can vary from those found in humans.

Coal-Mine Dust

The complex mixture known as coal-mine dust includes particles of coal of varied composition and cytotoxicity, as well as rock dust of varied silica content. Miners have been shown to have an excess of several chronic lung diseases, including coal workers' pneumoconiosis (CWP), progressive massive fibrosis (PMF), emphysema, and bronchitis. The largest incidence is that of CWP, as diagnosed by x-ray opacity, but the most serious with respect to the severity of effects and influence on mortality are emphysema, PMF, and bronchitis. There are two important limitations to our knowledge. First, the role of cigarette-smoking in the development of emphysema and bronchitis is not clear. Second, an acceptable animal model that might be used to provide insight into this problem is not available. Thus, our ability to use toxicologic approaches to develop a better understanding of the pathogenesis of the diseases of coal workers is severely limited.

REFERENCES

Albert, A. 1979. Selective Toxicity: The Physico-Chemical Basis of Therapy, 6th ed. John Wiley & Sons, New York. (662 pp.)

Albert, R. E., J. Berger, K. Sanborn, and M. Lippmann. 1974. Effects of cigarette smoke components on bronchial clearance in the donkey. Arch. Environ. Health 29:96–101.

Albert, R. E., H. T. Peterson, Jr., D. E. Bohning, and M. Lippmann. 1975. Short-term effects of cigarette smoking on bronchial clearance in humans. Arch. Environ. Health 30:361–367.

Ballantyne, B. 1985. Evaluation of hazards from mixtures of chemicals in the occupational environment. J. Occup. Med. 27:85–94.

Bandiera, S., K. Farrell, G. Mason, M. Kelley, M. Romkes, R. Bannister, and S. Safe. 1984. Comparative toxicities of the polychlorinated dibenzofuran and polychlorinated biphenyl mixtures which persist in Yusho victims. Chemosphere 13:507–512.

Berry, G., M. L. Newhouse, and M. Turok. 1972. Combined effect of asbestos exposure and smoking on mortality from lung cancer in factory workers. Lancet 2:476–478.

Bloch, W. N., Jr., T. R. Lewis, K. A. Busch, J. G. Orthoefer, and J. F. Stara. 1972. Cardiovascular status of female beagles exposed to air pollutants. Arch. Environ. Health 24:342–353.

Bohning, D. E., H. L. Atkins, and S. H. Cohn. 1982. Long-term particle clearance in man: Normal and impaired. Ann. Occup. Hyg. 26:259–271.

Calabrese, E. J. 1983. Principles of Animal Extrapolation. John Wiley & Sons, New York. (603 pp.)

Calabrese, E. J. 1986. Animal extrapolation and the challenge of human heterogeneity. J. Pharm. Sci. 75:1041–1046.

DeCaprio, A. P., D. N. McMartin, P. W. O'Keefe, R. Rej, J. B. Silkworth, and L. S. Kaminsky. 1986. Subchronic oral toxicity of 2,3,7,8-tetrachlorodibenzo-*p*-dioxin in the guinea pig: Comparisons with a PCB-containing transformer fluid pyrolysate. Fundam. Appl. Toxicol. 6:454–463.

Dube, M. F., and C. R. Green. 1982. Methods of collection of smoke for analytical purposes. Rec. Adv. Tobacco Sci. 8:42–102.

Faoro, R. B. 1975. Trends in concentrations of benzene soluble suspended particulate fraction and benzo(a)pyrene. J. Air Pollut. Control Assoc. 25:638–640.

Ferin, J., and L. J. Leach. 1973. The effect of SO_2 on lung clearance of TiO_2 particles in rats. Am. Ind. Hyg. Assoc. J. 34:260–263.

Freireich, E. J., E. A. Gehan, D. P. Rall, L. H. Schmidt, and H. E Skipper. 1966. Quantitative comparison of toxicity of anticancer agents in mouse, rat, hamster, dog, monkey, and man. Cancer Chemother. Rep. 50:219–244.

Hammond, E. C. and I. J. Selikoff. 1973. Relation of cigarette smoking to risk of death of asbestos-associated disease among insulation workers in the United States, pp. 312–317. In P. Bogovski, J. C. Gilson, V. Timbrell, and J. C. Wagner (eds.). Biological Effects of Asbestos. IARC Scientific Publication No. 8. International Agency for Research on Cancer, Lyon, France.

Hammond, E. C., I. J. Selikoff, P. L. Lawther, and H. Seidman. 1976. Inhalation of benzpyrene and cancer in man. Ann. N.Y. Acad. Sci. 271:116–124.

Hernberg, S. 1986. Uses of epidemiology in occupational health, pp. 317–340. In M. Karvonen and M. I. Mikheev (eds.). Epidemiology of Occupational Health. (WHO Regional Publications, European Series No. 20). WHO Regional Office for Europe, Copenhagen.

Hollander, D. 1981. Intestinal absorption of vitamins A, E, D, and K. J. Lab. Clin. Med. 97:449–462.

Hyde, D., J. Orthoefer, D. Dungworth, W. Tyler, R. Carter, and H. Lum. 1978. Morphometric and morphologic evaluation of pulmonary lesions in beagle dogs chronically exposed to high ambient levels of air pollutants. Lab. Invest. 38:455–469.

IARC (International Agency for Research on Cancer). 1981. IARC Monographs on the Evaluation of the Carcinogenic Risk of Chemicals to Humans, Vol. 25. Wood, Leather and Some Associated Industries. International Agency for Research on Cancer, Lyon, France.

IARC. 1982. IARC Monographs on the Evaluation of the Carcinogenic Risk of Chemicals to Humans, Suppl. 4. Chemicals, Industrial Processes and Industries Associated with Cancer in Humans (IARC Monographs, Volumes 1 to 29). International Agency for Research on Cancer, Lyon, France. (292 pp.)

IARC. 1983. IARC Monographs on the Evaluation of the Carcinogenic Risk of Chemicals to Humans, Vol. 28. The Rubber Industry. International Agency for Research on Cancer, Lyon, France. (486 pp.)

IARC. 1984. IARC Monographs on the Evaluation of the Carcinogenic Risk of Chemicals to Humans, Vol. 34. Polynuclear Aromatic Compounds, Part 3, Industrial Exposures in Aluminium Production, Coal Gasification, Coke Production, and Iron and Steel Founding. International Agency for Research on Cancer, Lyon, France.

IARC. 1986. IARC Monographs on the Evaluation of the Carcinogenic Risk of Chemicals to Humans, Vol. 38. Tobacco Smoking. International Agency for Research on Cancer, Lyon, France. (421 pp.)

ICRP (International Commission on Radiological Protection), Task Group on Lung Dynamics. 1966. Deposition and retention models for internal dosimetry of the human respiratory tract. Health Phys. 12:173–207.

Inskip, H., V. Beral, P. Fraser, M. Booth, D. Coleman, and A. Brown. 1987. Further assessment of the effects of occupational radiation exposure in the United Kingdom Atomic Energy Authority mortality study. Br. J. Ind. Med. 44:149–160.

Ip, C., and D. K. Sinha. 1981. Enhancement of mammary tumorigenesis by dietary selenium deficiency in rats with a high polyunsaturated fat intake. Cancer Res. 41:31–34.

Jammet, H., J. Lafuma, J. C. Nenot, M. Chameaud, M. Perreau, M. le Bouffant, M. Lefevre, and M. Martin. 1970. Lung clearance: Silicosis and anthracosis, pp. 435–437. In H. A. Shapiro (ed.). Pneumoconiosis: Proceedings of the International Conference, Johannesburg 1969. Oxford University Press, New York.

Karstadt, M., and R. Bobal. 1982. Availability of epidemiologic data on humans exposed to animal carcinogens: 2. Chemical uses and production volume. Teratog. Carcinog. Mutagen. 2:151–168.

Kazantzis, G., and J. C. McDonald. 1986. Work, health, and disease, pp. 43–67. In M. Karvonen and M. I. Mikheev (eds.). Epidemiology of Occupational Health. (WHO Regional Publications, European Series No. 20) WHO Regional Office for Europe, Copenhagen.

Klaasen, C. D. 1986. Distribution, excretion, and absorption of toxicants, pp. 33–63. In C. D. Klaasen, M. O. Amdur, and J. Doull (eds.). Casarett and Doull's Toxicology: The Basic Science of Poisons, 3rd ed. Macmillan, New York.

Koppanyi, T., and M. A. Avery. 1966. Species differences and the clinical trial of new drugs: A review. Clin. Pharmacol. Ther. 7:250–270.

Kotin, P., and H. L. Falk. 1960. The role and action of environmental agents in the pathogenesis of lung cancer. II. Cigarette smoke. Cancer 13:250–262.

Lawther, P. J., B. T. Commins, and R. E. Waller. 1965. A study of the concentrations of polycyclic aromatic hydrocarbons in gas works retort houses. Br. J. Ind. Med. 22:13–20.

Lippmann, M., R. B. Schlesinger, G. Leikauf, D. Spektor, and R. E. Albert. 1982. Effects of sulphuric acid aerosols on respiratory tract airways. Ann. Occup. Hyg. 26:677–690.

Little, J. B., E. P. Radford, Jr., H. L. McCombs, and V. R. Hunt. 1965. Distribution of polonium^{210} in pulmonary tissues of cigarette smokers. New Engl. J. Med. 273:1343–1351.

Logan, W. P. D. 1982. Cancer Mortality by Occupation and Social Class, 1851–1971. (IARC Scientific Publication No. 36; Studies on Medical and Population Subjects No. 44). H. M. Stationery Office, London, and International Agency for Research on Cancer, Lyon. (253 pp.)

McConnell, E. E., G. W. Lucier, R. C. Rumbaugh, P. W. Albro, D. J. Harvan, J. R. Hass, and M. W. Harris. 1984. Dioxin in soil: Bioavailability after ingestion by rats and guinea pigs. Science 223:1077–1079.

McCoy, G. D., S. S. Hecht, and E. L. Wynder. 1980. The roles of tobacco, alcohol, and diet in the etiology of upper alimentary and respiratory tract cancers. Prev. Med. 9:622–629.

Milham, S., Jr. 1983. Occupational Mortality in Washington State, 1950–1979. DHHS (NIOSH) Publication No. 83-116. U.S. Government Printing Office, Washington, D.C.

Milham, S., Jr. 1985. Improving occupational standardized mortality ratio analysis by social class stratification. Am. J. Epidemiol. 121:472–475.

Morgan, K. T., D. L. Patterson, and E. A. Gross. 1984. Frog palate mucociliary apparatus: Structure, function, and response to formaldehyde gas. Fund. Appl. Toxicol. 4:58–68.

NRC (National Research Council). 1975. Pest Control: An Assessment of Present and Alternative Technologies. Vol. I.: Contemporary Pest Control Practices and Prospects. National Academy of Sciences, Washington, D.C. (506 pp.)

NRC. 1978. Sulfur Oxides. National Academy Press, Washington, D.C. (322 pp.)

NRC. 1981. Indoor Pollutants. National Academy Press, Washington, D.C. (537 pp.)

NRC. 1983a. Diet, Nutrition, and Cancer. National Academy Press, Washington, D.C. (449 pp.)

NRC. 1983b. Polycyclic Aromatic Hydrocarbons: Evaluation of Sources and Effects. National Academy Press, Washington, D.C. (460 pp.)

NRC. 1984. Asbestiform Fibers: Nonoccupational Health Risks. National Academy Press, Washington, D.C. (334 pp.)

NRC. 1985. Epidemiology and Air Pollution. National Academy Press, Washington, D.C. (224 pp.)

NRC. 1986a. Environmental Tobacco Smoke: Measuring Exposures and Assessing Health Effects. National Academy Press, Washington, D.C. (337 pp.)

NRC. 1986b. Fire & Smoke: Understanding the Hazards. National Academy Press, Washington, D.C. (156 pp.)

NRC. In press. Disinfectants and Disinfectant By-Products. Drinking Water and Health, Vol. 7. National Academy Press, Washington, D.C.

NRC, Committee on Medical and Biologic Effects of Environmental Pollutants. 1977. Nitrogen Oxides. National Academy of Sciences, Washington, D.C. (333 pp.)

Office of Population Censuses and Surveys. 1978. Occupational Mortality. Decennial Supplement, England and Wales, 1970–1972. H. M. Stationery Office, London.

Pershagen, G., Z. Hrubec, and C. Svensson. 1987. Passive smoking and lung cancer in Swedish women. Am. J. Epidemiol. 125:17–24.

Phalen, R. F., J. L. Kenoyer, T. T. Crocker, and T. R. McClure. 1980. Effects of sulfate aerosols in combination with ozone on elimination of tracer particles inhaled by rats. J. Toxicol. Environ. Health 6:797–810.

Poiger, H., and C. Schlatter. 1980. Influence of solvents and adsorbents on dermal and intestinal absorption of TCDD. Food Cosmetics Toxicol. 18:477–481.

Radford, E. P. 1986. Ionizing radiation, pp. 726–733. In J. M. Last (ed.). Maxcy-Rosenau Public Health and Preventive Medicine, 12th ed. Appleton-Century-Crofts, Norwalk, Conn.

Rajewsky, B., and W. Stahlhofen. 1966. Polonium-210 activity in the lungs of cigarette smokers. Nature 209:1312–1313.

Rall, D. P. 1965. Conference on obstacles to the control of acute leukemia. Experimental studies of the blood-brain barrier. Cancer Res. 25:1572–1577.

Rall, D. P. 1971. Drug entry into brain and cerebrospinal fluid, pp. 240–248. In B. B. Brodie and J. R. Gillette (eds.). Handbuch der Experimentellen Pharmakologie, Vol. 28, Pt. 1. Springer-Verlag, Berlin.

Rall, D. P., M. D. Hogan, J. E. Huff, B. A. Schwetz, and R. W. Tennant. 1987. Alternatives to using human experience in assessing health risks. Ann. Rev. Public Health 8:355–385.

Rappe, C., S. Marklund, L. O. Kjeller, and M. Tysklind. 1986. PCDDs and PCDFs in emissions from various incinerators. Chemosphere 15:1213–1217.

Sawicki, E. 1967. Airborne carcinogens and allied compounds. Arch. Environ. Health 14:46–53.

Scharding, N. N., and F. W. Oehme. 1973. The use of animal models for comparative studies of lead poisoning. Clin. Toxicol. 6:419–424.

Schlesinger, R. B. 1986. The effects of inhaled acid aerosols on lung defenses, pp. 617–635. In S. D. Lee, T. Schneider, L. D. Grant, and P. J. Verkerk (eds.). Aerosols: Research, Risk Assessment and Control Strategies. Proceedings of the Second U.S.—Dutch International Symposium, Williamsburg, Virginia, The United States, May 19–25, 1985. Lewis Publishers, Chelsea, Mich.

Schlesinger, R. B., B. D. Naumann, and L. C. Chen. 1983. Physiological and histological alterations in the bronchial mucociliary clearance system of rabbits following intermittent oral or nasal inhalation of sulfuric acid mist. J. Toxicol. Environ. Health 12:441–465.

Selikoff, I. J., E. C. Hammond, and J. Churg. 1968. Asbestos exposure, smoking, and neoplasia. J. Am. Med. Assoc. 204:106–112.

Smith, R. L. 1973. Species variations in biliary excretion, pp. 76–93. In R. L. Smith (ed.). The Excretory Function of Bile. John Wiley & Sons, New York.

Speizer, F. E. 1983. Assessment of the epidemiologic data relating lung cancer to air pollution. Environ. Health Perspect. 47:33–42.

Umbreit, T. H., E. J. Hesse, and M. A. Gallo. 1986a. Comparative toxicity of TCDD-contaminated soil from Times Beach, Missouri, and Newark, New Jersey. Chemosphere 15:2121–2124.

Umbreit, T. H., E. J. Hesse, and M. A. Gallo. 1986b. Bioavailability of dioxin in soil from a 2,4,5-T manufacturing site. Science 232:497–499.

U.S. EPA (Environmental Protection Agency). 1982. Air Quality Criteria for Particulate Matter and Sulfur Oxides. EPA-600/8-82-029c. U.S. Environmental Protection Agency, Research Triangle Park, N.C. (3 vols.)

U.S. EPA. 1985. Guidelines for Preparing Environmental and Waste Samples for Mutagenicity (Ames) Testing. EPA 600/4-85-058. Environmental Monitoring System Laboratory, Las Vegas, Nevada. (Available from NTIS as PB 85-120144.) (255 pp.)

U. S. EPA. 1986. National Primary and Secondary Ambient Air Quality Standards. Code of Federal Regulations, Title 40, Part 50.

U.S. Surgeon General. 1982. The Health Consequences of Smoking: Cancer. DHHS (PHS) 82-50179. U.S. Public Health Service, Office on Smoking and Health, Rockville, Md. (322 pp.)

U.S. Surgeon General. 1983. The Health Consequences of Smoking: Cardiovascular Disease. DHHS

(PHS) 84-50204. U.S. Public Health Service, Office on Smoking and Health, Rockville, Md. (384 pp.)

U.S. Surgeon General. 1984. The Health Consequences of Smoking: Chronic Obstructive Lung Disease. DHHS (PHS) 84-50205. U.S. Public Health Service, Office on Smoking and Health, Rockville, Md. (545 pp.)

Vainio, H., K. Hemminki, and J.Wilbourn. 1985. Data on the carcinogenicity of chemicals in the *IARC Monographs* programme. Carcinogenesis 6:1653–1665.

Van der Berg, M., E. de Vroom, M. van Greevenbrook, K. Olie, and O. Huzinger. 1985. Bioavailability of PCDDs and PCDFs adsorbed on fly ash in rat, guinea pig, and syrian golden hamster. Chemosphere 14:865–869.

Way, E. L. 1967. Brain uptake of morphine: Pharmacologic implications. Fed. Proc. 26:1115–1118.

Weiss, B. 1978. The behavioral toxicology of metals. Fed. Proc. Fed. Am. Soc. Exp. Biol. 37:22–27.

WHO (World Health Organization). 1981. Health Effects of Combined Exposures in the Work Environment. Technical Report Series 662. WHO, Geneva. (76 pp.)

Wolff, R. K., M. Dolovich, G. Obminski, and M. T. Newhouse. 1977. Effect of sulphur dioxide on tracheobronchial clearance at rest and during exercise, pp. 321–332. In W. H. Walton (ed.). Inhaled Particles. IV. Part 1. Pergamon, New York.

3

Testing Strategies and Methods

Through the analysis of a series of published and unpublished studies, the committee found that two key steps must be taken for a problem with complex mixtures to be manageable. The steps are related to what questions are to be answered and what is known about the characterization of potential exposures. Another factor, related to those two steps, that is of crucial importance in selecting the appropriate testing strategy is whether the mixture under study is a known entity of expected uniformity and formulation or an unknown mixture of varied origins, such as leachates and runoff.

This chapter examines problem definition and choice of testing strategies from the perspectives of potential biologic effects of complex mixtures, the agents present in the mixtures, and the predictability of the data on one mixture with respect to other, related mixtures. Throughout, there is a concern for exposure potential. Without exposure, there is obviously no risk and therefore no reason for testing. The nature, extent, duration, and frequency of actual or anticipated exposure can all influence the selection of testing strategies and protocols. In particular, to the extent that any test protocol attempts to model the real world, extensive exposure data must be available in the study-design phase. (Chapter 2 deals specifically with such exposure-dose considerations.)

The definition of the problem will dictate the overall strategy to be followed. Reasonably standardized techniques that were developed for the testing of single chemicals can usually be adapted to study complex mixtures; only rarely must new approaches to toxicity testing be sought. The possibility of failing to observe adverse health affects or toxicity end points must be kept in mind, but this will probably occur no more often with complex mixtures than in the testing of single chemicals.

Predictability is an important factor in determining the approach to be used in testing mixtures. Stated most simply, it is of considerable value if, in the case of a series of mixtures, the testing strategy might provide information not only on the specific sample tested, but also on other, related mixtures. This potential use of data from one sample as a basis for prediction in other samples is far more difficult to resolve in the case of complex mixtures than for single agents, because of the possibility that interactions will influence physicochemical properties, bioavailability, biotransformation, and related phenomena.

Successful development of strategies for testing complex mixtures requires careful structuring of the questions to be asked, decisions regarding the order in which they are asked with respect to effects and composition of the mixture, and identification of the intended use of the information to be obtained. If the questions asked are not defined properly, the testing strategy required cannot be developed, nor can one expect to obtain results that are applicable to potential adverse health effects in humans. The major difference between mixtures and single chemicals is not in the testing strategies chosen, but in the importance of problem definition before testing.

PROBLEM AND QUESTION DEFINITION

The evaluation of the toxicity of a complex mixture should be preceded by understanding of the problem and definition of the questions to be answered. The extent and nature of testing should be guided closely by knowledge of what is known and what is to be learned. This section outlines the types of complex-mixture problems encountered and some questions that may be asked before selection of a strategy.

QUESTIONS RELATED TO EFFECTS

The type of exposure can also influence the selection of a testing strategy and specifically will affect the experimental design. (See Chapter 2 for a discussion of the relation between administered dose, or exposure, and biologically effective dose.)

The nature and magnitude of toxic effects dictate a number of questions:

- Under the conditions of expected human exposure, is the mixture potentially hazardous? Experimental designs should consider the routes, extents, and durations of exposure. Attention needs to focus on the extrapolation from experimental exposure to expected human exposure.
- If a toxic effect is observed epidemiologically, can a particular complex mixture be identified as responsible for the effect?
- Given an observable toxic effect in humans, can animal models be used appropriately to elicit it?
- How toxic is the mixture, compared with others under consideration? This question is not peculiar to complex mixtures, of course, but it is routine in

the assessment of the comparative toxicity of single chemicals in relation to related agents.

- What are the yardsticks of concern in preventing adverse health effects in the exposed general population? Should a spectrum of toxicity tests in a standardized format be done, or should a specific end point—for example, mutagenesis, teratogenesis, immunotoxicity, or behavioral aberrations—be the focus of testing? In retrospective epidemiologic studies, these responses tend to be subtle and inconclusive, the relationship between effect and causative agent being tenuous at best. Whatever end points are chosen must be considered in relation to human exposure to a given complex mixture and a set of plausible circumstances of exposure.

Questions Related to Causative Agents

Questions related to causative agents are peculiar to situations in which a toxic effect has been observed but the causative agents are unknown. They arise for such reasons as a desire to understand mechanisms of toxicity and ways to reduce toxicity. The questions include the following:

- What is the source of the toxicity of the complex mixture? The effects might be associated with a component that is active because it constitutes a major proportion of the mixture, is a minor constituent with high intrinsic toxicity, or is the component with the greatest bioavailability (due to physicochemical properties, etc.); or the effect might be a property of a combination of components.
- Is the chemical composition of the mixture known? Knowledge of the toxic potential of mixture constituents might permit the development of strategies to examine toxic end points of special interest.
- Can the toxicity of a product be reduced by altering its composition? This question generally requires an identification of the causative agents, if one is to determine whether removal of such components is both possible and likely to reduce toxicity.

Questions Related to Predictability

Questions related to predictability arise under two circumstances: when chemical constituents of a mixture are known and effects of the mixture need to be predicted, or when both agents and effects are known for one situation but need to be predicted for another. The following are examples of predictability questions:

- How might interactions of constituents contribute to the toxicity of a mixture at different extents and durations of exposure?
- Are the measured biologic effects reproducible with different samples of the mixture? A positive answer would suggest that one sample would suffice to

characterize the predominant toxicity attributable to the mixture. A negative answer should lead one to query the stability or heterogeneity of the mixture and address problems of sampling.

• Are the toxicity results useful for predicting potential toxicity of similar mixtures? Many complex mixtures are heterogeneous, their compositions varying both qualitatively and quantitatively from source to source and being influenced by the conditions governing their formation. For example, is each municipal waste incinerator unique in its emissions, or can conclusions be drawn about incinerators in general by sampling only a few sources? If so, how many samples are needed regarding the range of burn conditions?

STRATEGIES

Once the appropriate questions have been formulated, the strategies to answer them must be considered.

The strategies for evaluating complex mixtures have been organized into three categories, according to the kinds of questions to be answered—those related to effects, to causative agents, and to predictability. Although the strategies are often applied in combination to address a series of questions, they are described separately here to exemplify the specific approaches that can be used. The integration of strategies is discussed later in this chapter.

Strategies Related to Effects

Biologic effects of complex mixtures can be the same as those of single agents. There is no compelling reason to expect their effects to be different, and the investigator will normally follow the same patterns of observation and recording used for single agents. If the mixture contains representatives of a wide range of chemical types and a test involves intact animals, the mixture might be expected to affect many organ systems.

As Chapter 2 notes, documented dose-response relationships in humans for exposures to most mixtures are relatively rare. Strategies for testing mixtures should avoid overcomplex programs and take into account the normal dose rate and frequencies of exposure. However, testing strategies must always be sensitive to unpredicted outcomes.

Effect-related strategies generally involve well-established stepwise toxicologic evaluations of a test substance—in this case a mixture (NRC, 1977). They are concerned initially not with causative agents, but with observable outcomes (effects). This approach uses the complete material, and increasingly elaborate studies are based on the outcome of early, simpler studies. Causation may be inferred from thoughtful evaluation of experimental observations. Such factors as time of onset of a toxic response, duration of effect, reversibility of effect, physiologic system involved, organ pathology, and possible links between affected organs can all be parts of a pattern attributable to specific agents or classes of agents. Examples would include cholinesterase inhibitors, heavy metals, and hydrocarbon-based solvents. In the case of some mixtures, identification of causative agents might not even be the information desired. The main interest might be to confirm or eliminate a causal relationship between an observed effect and an uncharacterized material. In such instances, it is proper to test the sample mixture without conducting characterization studies.

An investigator has a number of design options that avoid premature commitment to an extensive slate of protocols. Several options are described below.

Tier Testing

Tier testing may be likened to a series of sieves with smaller and smaller pores (NRC, 1984). At each stage, some findings require followup and others are considered adequate or conclusive. Figure 3-1 illustrates a form of tier

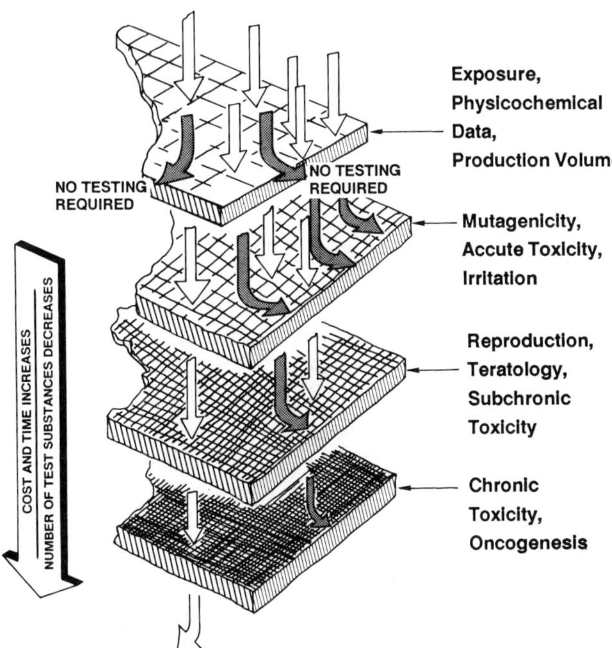

FIGURE 3-1 Schematic diagram of tier testing where the first tiers consist of less-expensive and less-time-consuming screening tests and progress at higher times and more-expensive and time-consuming tests.

testing used in the study of commercial mixtures; predetermined triggers (or end points) dictate whether the next stage is required. A comparable system is a part of the premarketing notification system of the European Economic Community; in this instance, the triggers are related to the amounts of material intended to be produced annually. Not unexpectedly, the cost and time required for each tier are greater than those for the preceding one.

A tier-testing approach has been applied to a number of complex mixtures (e.g., diesel emissions, synthetic fuels, and concentrated organic mixtures from drinking water) to identify potential toxic effects resulting from exposures to these mixtures.

Screening Studies

Screening studies are less integrated than tier-testing programs. The investigator first determines which biologic end points are of interest and then screens or gains preliminary information regarding those effects, rather than conducting definitive work for each effect. The simplest test that reliably points to an effect is chosen. The criterion should be sensitivity (few false negatives), rather than selectivity (few false positives). Assays are done sequentially until the investigator is able to rank the test material with respect to effects of interest. Where necessary, more definitive studies are done on agents about which there is uncertainty, whose exposure potential is greatest, about which there is substantial social concern, or whose potency or other biologic considerations warrant further study. The carcinogenicity studies initially performed by the National Cancer Institute's Bioassay Program relied on a screening program designed to identify potential carcinogenic agents (NCI, 1976). Agents were selected generally because of widespread use or because of suspicion of potential carcinogenicity. Doses were high, and the result was a measure of carcinogenic potential (not risk or hazard). Current National Toxicology Program long-term carcinogenicity bioassays have been modified to provide more information (e.g., by using more than two doses and collecting pharmacokinetic data) (Huff, 1982).

Screening studies have been successfully applied to mixture problems, such as the hexacarbon-neuropathy case. These studies revealed that an organized, scientific approach to problem-solving has a high probability of success, even with complex mixtures, where there is a combination of focused questions, assay methods, and good luck. The focused questions (referred to previously) resulted in the isolation of the problem to specific work areas that used materials (mixtures) of known or knowable composition. The same focused questions went beyond a general description of the disease to its specific characteristics and permitted the establishment of links to other case reports. The availability of suitable assay systems allowed the screening of hundreds of workers for neurologic deficits and permitted the concerns about the workplace to be concentrated in the print department. The assays also allowed some agents to be eliminated as causes, because the neurologic findings were not of the expected kind. An animal assay allowed mixtures or individual chemicals to be screened for the neurologic disorder and for structure-activity studies to be conducted. (For additional information on this case, see Appendix C.)

Matrix Testing

Matrix testing involves the systematic manipulation of several variables to define the "universe" of materials and outcomes. Once critical variables have been identified, the agents can be arranged as loci in a matrix field whose dimensions are defined by those variables, which are presented normally in a two-dimensional plot. Additional dimensions can be considered, although the possibilities for graphic representation diminish as dimensions are added.

An example of this approach may be found in a series of animal toxicology studies on hydrocarbon solvents published by Carpenter et al. (1975a–h, 1976a–e, 1977a-c, 1978). The underlying premise was that the toxicity of this class of materials was related to boiling range and aromaticity. The matrix is shown in Figure 3-2. Boiling range was a routine specification available for all such materials and is related to number of carbon atoms (and molecular weight). The aromaticity dimension reflected the existing literature, which shows that, for a comparable number of carbon atoms, toxicity decreases from aromatic hydrocarbons to naphthenic hydrocarbons to paraffinic hydrocarbons (Scala, in press).

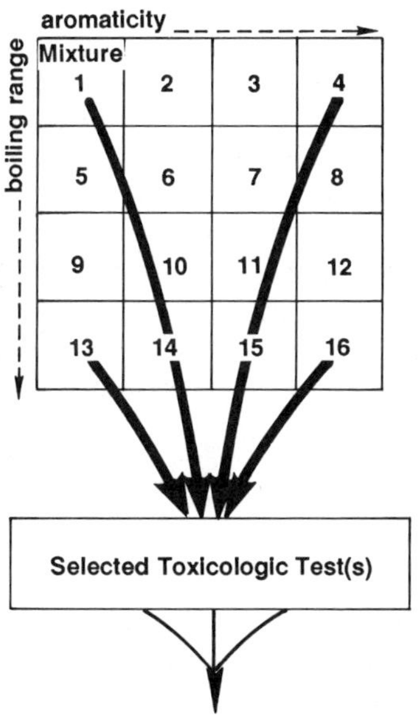

FIGURE 3-2 Example of matrix testing in which critical variables for series of mixtures are manipulated to define boundary conditions such that toxicologic testing of selected variables would result in predictability of effects within matrix.

The individual mixtures of commercial importance were placed on a matrix with these dimensions. Selected mixtures were tested in varied toxicologic screening paradigms. The results of each test dictated where in the matrix the next sample would be chosen. It was hoped that defining "boundary" conditions would make it possible to position effects within the matrix.

Battery Approach

In the battery approach, an array of bioassays is used. This strategy (Figure 3-3) has been used in the field of genetic toxicity, in which the emphasis is on determining whether agents are genotoxic, whether metabolic conversion is a prerequisite for activity, and what sort of mechanisms might be involved (Heussner et al., 1985). Just as a diner might construct a meal by going through several courses on a complex menu, so an investigator can provide answers to the specific questions posed by selecting tests from an available battery.

Comparative-Potency Approach

Comparative approaches in toxicology generally involve studies of the effects of one substance in different species or bioassays (as in the battery approach) or studies of the effects of a series of substances in one or a few bioas-

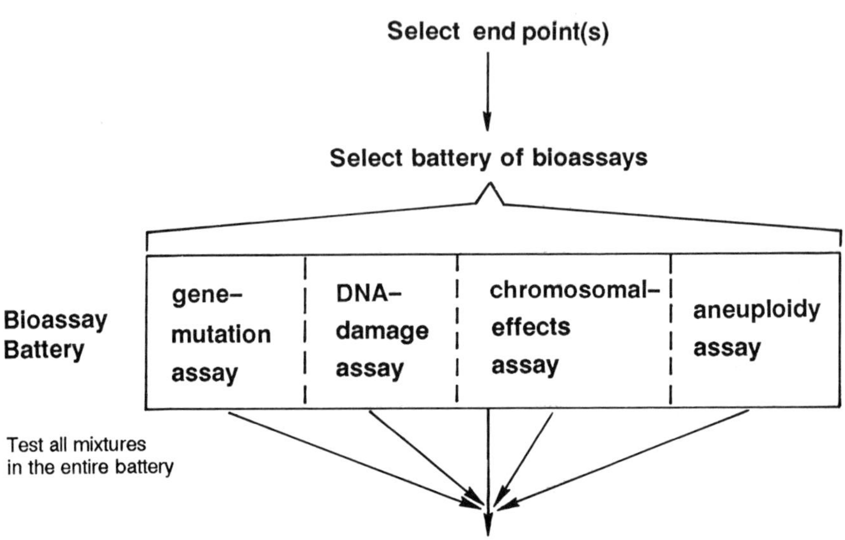

FIGURE 3-3 Example of battery approach to toxicologic testing of mixtures in which genetic-toxicity test battery is constructed to test for different genetic effects.

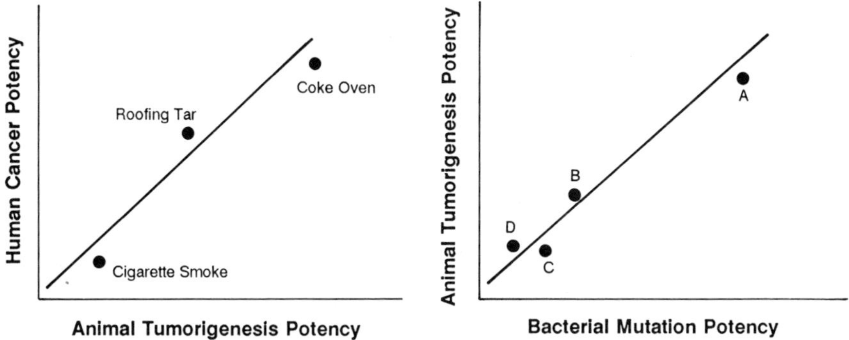

FIGURE 3-4 Example of comparative-potency approach to evaluating diesel emissions.

says. The latter is described here as the comparative-potency approach; it is particularly useful for the evaluation of complex mixtures (Figure 3-4).

Many mixture problems involve the grouping of mixtures, such as fuels derived from different sources (e.g., petroleum, shale, and coal), soots emitted from different combustion sources, hazardous wastes from different sites, effluents from different industries, emissions subject to different control technologies, and emissions from the burning of different products. The problem posed to the toxicologist might be whether one mixture is more or less toxic than another. All soots from incomplete combustion are expected to contain carcinogenic polycyclic organic compounds, so the toxicologist might have to address the question of whether a new combustion source will produce soot that is more potent in a carcinogenicity bioassay than soot from current sources.

A comparative-potency method for cancer risk assessment has been developed and tested with estimates of human lung-cancer risk (Albert et al., 1983). The data base used was associated with several mixtures, including emissions from coke ovens, roofing-tar pots, cigarette-smoking, and automotive engines (Lewtas, 1983; Nesnow et al., 1982a,b; Lewtas et al., 1983). The method was based on the hypothesis that relative potency of different carcinogens in different bioassay systems is constant (Lewtas et al., 1983). The mathematical expression for the constant-relative-potency model is as follows: (relative potency in bioassay 1)/(relative potency in bioassay 2) $= k$. This assumption of constancy is implicit in any comparison that uses the relative toxicity of two substances in animals to predict which one would be less toxic in humans. The assumption is testable if the relative potency of two mixtures or components can be determined in one bioassay (e.g., humans) and compared with the relative potency in a second bioassay. The hypothesis was tested for three complex organic emissions from a coke oven, a roofing-tar pot, and cigarettes by using human lung-cancer data from epidemiologic studies of humans exposed to these emissions (Albert et al., 1983) and testing the emissions in a series of short-term mutagenicity bioassays and animal tumorigenicity bioassays.

The comparative approach has been widely used in the petroleum industry. In one such effort, the comparative acute toxicity was determined for a series of 19 petroleum hydrocarbon products ranging from light oils and gasoline to heavy fuel oils. The testing included standard oral and dermal acute and sub-chronic toxicity tests, as well as a series of sensitization and irritation tests (Beck et al., 1982). Another example involved shale-derived fuels and other synfuels whose toxicity was determined primarily in a wide variety of comparative studies, including general toxicity, target-organ, behavioral, mutagenicity, carcinogenicity, teratogenicity, and neurotoxicity tests (see MacFarland et al., 1984, and Mehlman et al., 1984). These studies were not conducted specifically to provide a quantitative estimate of human risk associated with the mixtures, but rather to gauge the comparative toxicity potential of this series of mixtures.

The design considerations for comparative-potency studies include factors that might not always be appropriate in other strategies. Some examples are the following:

- *Simultaneous evaluation of all comparisons in one experiment*. This might be desirable if bioassay variability between experiments is large. It is not possible in many chronic bioassays involving large numbers of animals, but it is often the best approach in some in vitro bioassays. This approach has been recommended specifically for comparative studies with the *Salmonella typhimurium* reverse-mutation plate incorporation assay (Lewtas, 1983).
- *Exposure doses (or concentrations) needed for statistical analysis and potency measurement*. The use of identical measures of dose across all the mixtures being studied usually facilitates the statistical analysis, including estimation of potency. If the potency range of the mixtures being tested is very large, the dose ranges for the assay might not overlap. Range-finding studies are usually needed to determine doses. Comparative potency (e.g., effect per unit dose or dose causing an effect) must be based on an exposure dose or concentration and can be expressed in any terms normally used to describe or characterize the toxic effect. Various statistical methods are available for determining linear and nonlinear slopes, as well as the relative dose required to cause a specific effect (Lewtas et al., 1983).
- *Study objectives and integrated use of the data*. The objectives of a specific comparative-potency study and the expected use of the data are important in study design. Feder et al. (1984) have proposed a strategy for evaluating the toxicity of mixtures of gasoline blends that uses stagewise organization, fractional factorial designs, multiple bioassays, and a standardized reference fuel as a center point. (This "global" approach is discussed further in Chapter 5.) The purposes are to minimize the testing required to indicate the variability in

toxicity among members of a class of similar mixtures (e.g., blends of gasoline) and to provide a data base that will allow future predictions based on composition and response-surface modeling.

Strategies Related to Causative Agents

Bioassay-Directed Fractionation

The objective of bioassay-directed fractionation is to identify the biologically active (bioactive) components of mixtures.

The choice of bioassays depends on the mixture being tested and on what is known about its chemical composition or the toxic effects. Bioassays that have been used include assays of tumor initiation, tumor promotion, mutagenicity, cytotoxicity, target-organ effects, and 2,3,7,8-tetrachlorodibenzo-*p*-dioxin (TCDD) cellular receptor binding.

Bioassay-directed fractionation has been extensively used to identify biologically active components of cigarette smoke, particularly with cigarette-smoke condensate (CSC). Fractionation of CSC followed by mouse-skin carcinogenicity bioassays led to the identification of specific fractions of tobacco smoke that contain tumor-initiating activity and other fractions that contain tumor-promoting or carcinogenic activity (Bock et al., 1969; Hoffmann and Wynder, 1971). Later, short-term mutagenicity and cellular-transformation bioassays were used to identify the most biologically active fractions and in some cases to identify specific chemicals in those fractions that could account for a portion of the mutagenic or tumorigenic activity (DeMarini, 1983). These studies concluded, for example, that polycyclic aromatic hydrocarbons in the neutral fraction contribute significantly to the tumor-initiating activity of the neutral fraction of CSC, whereas amines, aza-arenes, and other nitrogen-containing compounds account for much of the mutagenic activity of the basic fraction of CSC (DeMarini, 1983; IARC, 1986). (For additional information on this case, see Appendix C.)

Nearly all the methods that have been described or are available for chemical analysis have been used to fractionate mixtures for bioassay. Often, several methods are used sequentially. Many of these approaches are described in Chapter 4, and the overall approach is diagrammed in Figure 3-5.

From the perspective of the toxicologist, there are several important points in selecting fractionation methods and conducting these studies, including the following:

- *Recovery of mass and bioactivity*. At each point in the fractionation scheme where bioactivity is measured, recovery of mass and bioactivity should be accounted for. If the sum of the bioactivities of the mass-weighted fractions equals that of the unfractionated mixture and that of a reconstituted mixture,

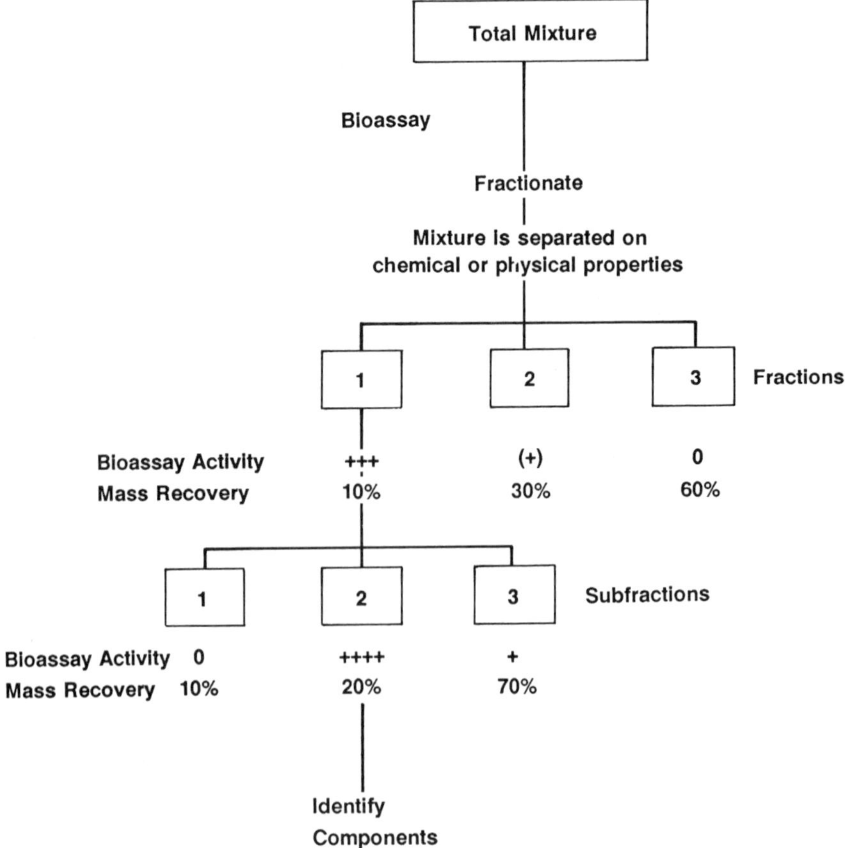

FIGURE 3-5 Diagrammatic representation of bioassay-directed fractionation where mixture is sequentially separated into fractions and fractions are bioassayed. Effort is concentrated on further fractionation and chemical characterization of biologically active fractions.

then it at least *appears* that all the bioactivity is accounted for. One should be cautious, however: simultaneous chemical changes as a result of the fractionation procedure could increase the bioactivity of some components and decrease that of others and thus result in the appearance of no change.

• *Identification of minor components*. In several reported cases, the bioactive components of mixtures were minor, according to their mass, but extremely bioactive. Examples of such components are the strongly mutagenic dinitropyrenes, found in xerographic toners (Rosenkranz et al., 1980) and diesel exhaust (Salmeen et al., 1984); the potent toxicant aflatoxin, found in peanuts (Dichter, 1984; Linsell, 1979); and the potent contaminants of polychlorinated biphenyls, such as polychlorinated dibenzofurans, found in emissions from burning transformers (Milby et al., 1985), and the bicyclophosphate ester found in the thermal decomposition products of a fire-retardant polyurethane foam (Petajan et al., 1975).

• *Quantitation of contribution of bioactive components*. Once a highly bioactive fraction is obtained and chemical analysis provides a list of the confirmed or suspected components, additional information and sample availability requirements must be anticipated. Bioactivity data on each of the compounds identified must be available. Pure standards must be obtained to assay the identified compounds and determine their contribution to the total activity observed, or fractionation and biologic testing must continue.

Those issues are described from the perspective of the analytic chemist in Chapter 4.

Pairing

Pairing (depicted in Figure 3-6) combines information obtained independently from the literature or computer data bases with information on chemicals known to be bioactive and chemicals identified analytically as present in the mixture of interest. This approach is used in literature reviews (Bridbord and French, 1978; NRC, 1983; Bingham et al., 1980).

Chemical Search

In a chemical search (as illustrated in Figure 3-7), bioassay data on pure chemicals and analytic data on the bioactive components of a mixture are used.

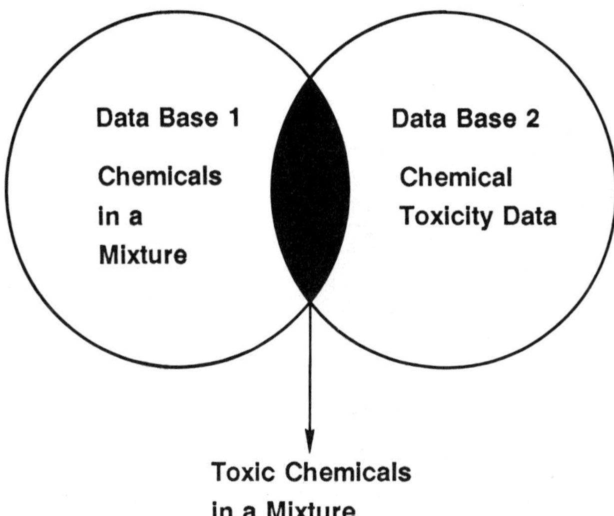

FIGURE 3-6 Two data bases can be paired or combined; overlap represents toxic chemicals identified in mixture. This method is described as pairing.

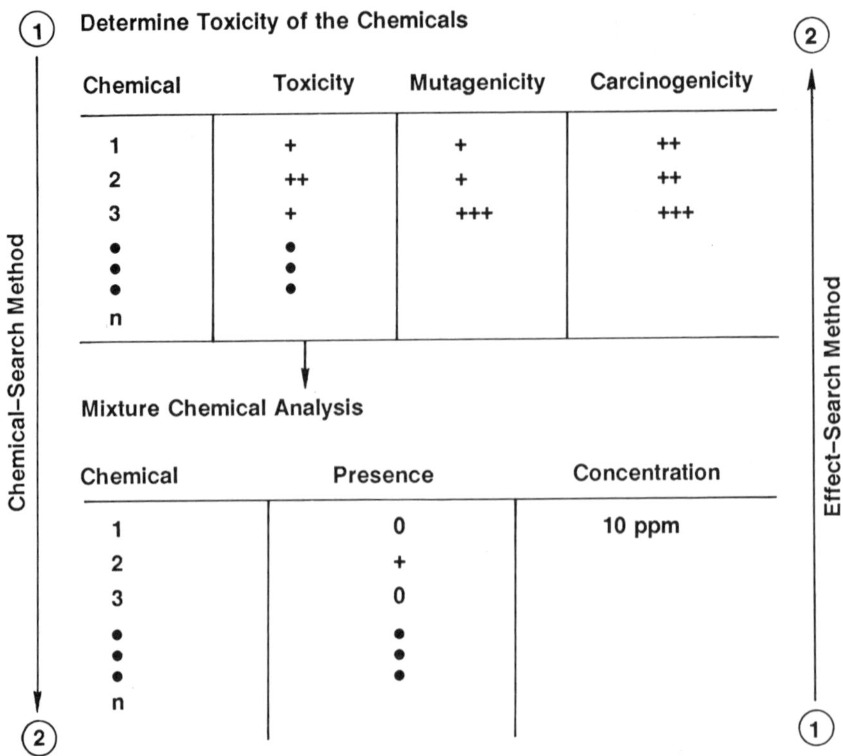

FIGURE 3-7 Illustration of chemical-search and effect-search methods for determining causative agents in mixture.

This method has been used with nitroparaffins (Hite and Skeggs, 1979) and with polycyclic aromatic hydrocarbons (Bos et al., 1984). It can be used to evaluate a series of similar mixtures rapidly after bioassay-directed fractionation has identified the class of compounds of interest, such as nitrated polycyclic aromatic hydrocarbons (Pederson and Siak, 1981). Later, it can be used to search chemically for the most mutagenic compounds in other related mixtures (Nishioka et al., 1985).

Effect Search

Effect search (Figure 3-7) is the reverse of chemical search. Initially, chemical methods are used to identify as many chemicals as possible in the mixture. That step is followed by bioassay studies or literature searches on the chemicals, in an attempt to identify those related to the effects of interest. The approach has been widely used in dealing with hazardous wastes (Houk and

Claxton, 1986; Andon et al., 1984) and was used by Florin et al. (1980) to identify hundreds of compounds found in tobacco smoke.

A combination of chemical-search and effect-search strategies has been used to identify acutely toxic agents in fires (Alarie and Anderson, 1979; Crane et al., 1977; Hilado, 1976; Klimisch et al., 1980; Levin et al., 1982; and Yusa, 1985). Concern about the toxicologically responsible single or multiple components in fire atmospheres has been approached through extensive chemical analyses (Levin, 1986). The prediction of acute toxic hazard, which includes many other components in addition to toxicity, has used a computer model that integrates many of the necessary pieces of the puzzle. The strategy to determine the toxicity factors needed for input into the hazard assessment has been to determine the individual and combined toxicities of a small select number of fire gases, predict the toxicity of the thermal decomposition products from materials based on the chemical analysis of the selected gases, and verify the prediction with a few animal exposures (Levin et al., in press). (For additional information on this case, see Appendix C.)

Bioassay Identification

Bioassay identification relies on the development of specialized bioassays that can detect chemicals of a particular class or structure. Examples of such assays are the TCDD receptor assay, monoclonal antibody assays, and genetically engineered bacterial strains that can metabolize, and thus mutate, only after exposure to particular compounds.

Strategies Related to Predictability and Models

The goal of strategies related to predictability is to predict the toxicity of materials related to a sample of known toxicity. In many instances, inadequacy of resources (such as time and sample availability) prevents the design and conduct of laboratory tests on a series of complex mixtures. Hazardous-waste sites, for example, present problems associated with temporal and spatial variability of leachate composition. It might be necessary to predict the risk connected to a time- and space-integrated sample that is representative of the probable human exposure. These strategies are best described as model-driven, because they are based on empirical, theoretical, statistical, or mechanistic models (described in greater detail in Chapter 5).

Testing Mixture Components

When the composition of a mixture is known and no biologic data are available on the whole mixture, its toxicity can be approximated by summing the known toxicities of the components (see Chapter 5 and Appendix D). A weakness of this approach is the likely lack of complete chemical composition data. Also, the strategy assumes that no interactions occur. That assumption can be tested biologically at several doses with interaction studies. Interaction studies have been performed with many chemicals. Usually, however, only two agents were examined; in a few, three were examined. The fire-toxicology case study illustrates the use of matrix testing to examine the interaction of mixture components. (For more information, see Appendix C.)

When two substances are involved in toxicity testing, their individual characteristics are investigated completely, as well as those of the combination. Usually, dose-response relationships are worked out in detail for one agent and then essentially repeated in the presence of the second agent (Klaasen, 1986). The purpose is to establish whether the presence of the second agent alters the dose-response profile of the first. The two agents can be used as a mixture or sequentially. The protocols involved in this type of strategy are very effect-dependent.

Essentially, the purpose of this strategy is to test the hypothesis that the effect seen is other than additive. Mitchell (1976) pointed out that, if both substances are active, the first step is to determine the potency of one relative to the potency of the other in effecting the response. The second step is to combine fractional doses of the substances and compare the results with the results of using standard doses of the substances individually. The null hypothesis is that the two substances will behave as though they are different forms of the same substance and hence produce additive effects.

The modification of the dose-response curve of a toxicant given in combination (concurrently or sequentially) with another substance can have important implications for prediction. These implications have been described for the potentiation of chloroform-induced liver injury by the pesticide chlordecone (Plaa and Hewitt, 1982a) and the potentiation of haloalkane-induced liver and kidney injury by ketones or ketogenic agents (Plaa and Hewitt, 1982b). If the dose-response curve is displaced to the left but is parallel to the curve obtained with animals that are not pretreated, the potentiating agent decreases the effective dose of the toxicant under study. If the curve does not shift laterally (effective dose is unchanged) but the slope of the curve increases, the response to the toxicant is exaggerated by the potentiating agent. If the curve is displaced to the left and the slope is increased, the potentiating agent lowers the effective dose of the toxicant and exaggerates the response to it.

The implications of each situation are different. With a parallel lateral displacement, the effective dose is smaller, but the effect threshold can be estimated rather accurately. If the curve has rotated, the effective dose is unaltered; the established no-observed-effect dose remains the same. If the curve has shifted laterally and the slope has changed, neither that dose nor the severity of response can be predicted from the responses observed in the absence of the second chemical. Chlordecone pretreatment in mice results in a leftward shift of the chloroform dose-response curve; in rats, however, the curve appears both to shift laterally and to rotate (Plaa and Hewitt, 1982a).

In selecting doses for interaction studies when both agents are active, it is necessary to decide the fraction of the effective dose of each agent that should be administered. If the dose-response curves for the two agents are parallel, equivalent fractions of the median effective doses or equipotent doses of the agents can be selected. However, if the curves are not parallel, one must use equipotent doses; equivalent fractions of the median effective doses can result in misinterpretation of the action of the combination (Clausing and Bieleke, 1980; Mitchell, 1966).

Dose-dependent interactions are not always monophasic linear relationships. In ketone potentiation of haloalkane-induced liver or kidney injury, biphasic phenomena can occur, with low doses of ketone potentiating and larger doses protecting (MacDonald et al., 1982; Hewitt and Plaa, 1983; Brown and Hewitt, 1984).

In the case of simple mixtures, knowledge about the mechanisms of action facilitates the designing of experiments. The more one knows of how the toxic response is initiated and propagated, the better the study. It is essential to understand the pharmacodynamic and pharmacokinetic characteristics of the agents under study.

Mechanistic Studies and Models for Interactions

An analysis of the mechanistic principles underlying the development of interactions has shown that only a few biologic phenomena can be mitigated or amplified through multichemical exposure (Witschi, 1982). Acute effects amplified by interaction most often include cell death in particular organs or tissues, and such an event can often be traced back to the increased formation of toxic compounds. Potential mechanisms of chronic effects of interactions include the formation of DNA adducts, prolongation of the life of free radicals, enhancement or impairment of transport across epithelial barriers, and induction of mixed-function oxidases. Components of the disease process and its secondary effects (hepatotoxicity, nephrotoxicity, and tumor promotion) can influence the development of a primary disease.

It might be appropriate to see whether mixtures or selected components affect the potential mechanisms for interaction. In many cases, that can be done in simple in vitro systems.

Empirical Models

The simplest experimental approach to estimating the effects of an unknown mixture is to derive an empirical model. The requirements for such a model are a biologic effect that can be quantified and one or more characteristics of the mixture that can be quantified (e.g., boiling point and solubility). It is not necessary to know the chemical composition of the mixture or the causative agent(s), but knowledge of the concentration of one or more chemical constituents could be useful.

The extent to which an observed effect is correlated with measured characteristics indicates the predictive utility of the model, although it does not imply causation or mechanism.

Aquatic Test Systems

Although this report focuses on mammalian toxicity testing, the committee recognizes that aquatic toxicity test systems are being used for hazard evaluation and risk assessment in a number of fields, particularly in chemical-mixture toxicity tests. The standard acute toxicity tests are still performed in a number of laboratories and for the most part yield the evaluations required by regulatory agencies.

In recent years, however, interest has shifted away from the standard tests, and more and more information concerning the biochemical aspects of aquatic species is becoming available. Along with the development of this data base, toxicologists are gaining new insight into how chemicals interact with various biochemical and physiologic pathways.

In the last few years, a number of field studies have attempted to correlate chemical exposure to the concentrations of xenobiotic metabolizing enzymes (the mixed-function oxidase system) in feral fish populations (Stegeman et al., 1986). In the laboratory, it has been shown that fish will respond to exposure to a number of chemicals by an increase in or induction of specific isozymes of the metabolizing system (Lech et al., 1982; Buhler and Rasmusson, 1968). Some of the enzymes of the rainbow trout and scup have been purified and cloned (Klotz et al., 1984; Williams and Buhler, 1983).

A number of fish species have been shown to be sensitive to various chemical carcinogens, both direct and indirect (e.g., diethylnitrosamine or methylazoxy methanol acetate). Exposure to these chemicals has resulted in rapid tumor formation (Stanton, 1965; Ishikawa et al., 1975) and unscheduled DNA synthesis (Ishikawa et al., 1978). Because of these types of responses, there is a great deal of interest in their use as tier II test species in the evaluation of chemicals and chemical mixtures as potential carcinogens.

INTEGRATION OF STRATEGIES

The strategy for toxicity testing of mixtures depends on at least three factors: complexity of the mixture, knowledge of the composition (constituents) and effects of the mixture, and the problem or question. The influence of each of these factors, individually and in combination, on testing strategy is examined below.

Complexity of Mixture

As the complexity of a mixture increases, the likelihood that its toxic components can be completely characterized decreases. The optimal testing strategy for the simplest two-component mixture is to test the components in such a way as to determine whether combining them causes interactions. Toxicologic testing of components with a design that elucidates interactions and their mechanisms will provide data that can be useful in models for predicting the toxicity of other simple mixtures that contain the same components.

For combinations of more than two constituents, the task of assessing the interactions of all components becomes overwhelming and impractical, because the number of binary combinations equals $(n^2 - n)/2$. Environmental mixtures are therefore evaluated by administering either the total mixture or a few fractions or ingredients. If the latter approach is selected, efficient experimental designs can be extracted from the statistical literature on what are called fractional factorial designs. These are discussed in Chapter 5. Figure 3-8 describes a design for the preliminary assessment of 16 substances with 24 mice. Note the sequences within each group and the fact that the number of animals

		A GROUP $S_1S_2S_3S_4$	B GROUP $S_5S_6S_7S_8$	C GROUP $S_9S_{10}S_{11}S_{12}$
α GROUP	S_1 S_5 S_9 S_{13}	2 MICE	2 MICE	2 MICE
β GROUP	S_2 S_6 S_{10} S_{14}	2 MICE	2 MICE	2 MICE
γ GROUP	S_3 S_7 S_{11} S_{15}	2 MICE	2 MICE	2 MICE
δ GROUP	S_4 S_8 S_{12} S_{16}	2 MICE	2 MICE	2 MICE

FIGURE 3-8 A fractional factorial design for the preliminary assessment of 16 substances with 24 mice. The intersection of the β group and the C group is cross-hatched to indicate a toxic response in one of the animals in that cell.

exposed to each pair of substances is at least double the number in each cell. The intersection of the β group and the C group is cross-hatched to indicate a toxic response in one of the animals in that cell. To identify the particular substances responsible for that response, the seven substances assigned to the cell could be divided into another fractional factorial design or tested in pairs individually.

Knowledge of Constituents and Effects

The selection of strategies and their application to toxicologic testing are often to a great extent either agent-driven or effect-driven. For example, if a very complex mixture is known to be carcinogenic and mutagenic, the driving force will often be the identification of agents (or components) that are causing the effect. That will lead to a bioassay-directed fractionation strategy. The research published earlier on unleaded-gasoline vapors and diesel particle emissions has been driven by the observed carcinogenic effect (MacFarland et al., 1984).

When an agent or class of agents, such as TCDD or polychlorinated biphenyls (PCBs), is known to be present in a mixture, the testing will often be driven to elucidate the effects or the influence of other components on an identified effect. Knowledge of components or effects has direct influence on the analytic and toxicologic testing strategy. When neither the effects nor the constituents are known, there is a driving force to elucidate both, and the strategy will be influenced more by the specific problems or questions being asked about the mixture.

Problems or Questions in Relation to Strategies

Problem definition and the formulation of specific questions are important steps that must be taken before selection of a testing strategy. Examples of the types of complex-mixture questions and proposed strategies were discussed early in this chapter. They can be grouped into questions regarding effects and risk, causative agents, and predictive value.

The questions usually arise sequentially, so questions about the effects of a mixture would logically be raised before questions on causative agents. Questions on predictive value, however, have been raised when both a mixture's effects and agents were known and when very little was known about it other than what some of its constituents were. It is more productive to address the question of predictive value with interaction studies when effects and agents are known. When neither the effects nor all the agents are known, the strategy should be to elucidate the agents and effects before attempting to address how

TABLE 3-1 Relationship Between Questions/Problems and Associated Strategies

Questions	Strategies
Effects and risks	Toxicologic evaluation of mixtures
	Tier/screening
	Matrix/comparative
Causative agents	Bioassay-directed fractionation
Predictive value	Toxicologic evaluation of components

and whether predictions could be made from knowledge of a limited number of constituents. The strategies can clearly be grouped according to the general type of questions they address, as shown in Table 3-1.

Formulation of Overall Strategy

Each of the three factors above needs to be considered in selecting the final strategy for toxicologic studies (Figure 3-9). Most research on mixtures involves a logical progression of strategies as knowledge is gained about effects and agents. As new concerns are raised over exposures to mixtures, the factors discussed above should be considered in selecting strategies to provide data useful in answering questions about potential human risk and in selecting steps to protect public health. Figure 3-10 diagrams a hypothetical flow of a process that would consider each of the above factors and use the strategies presented here to address specific problems and questions.

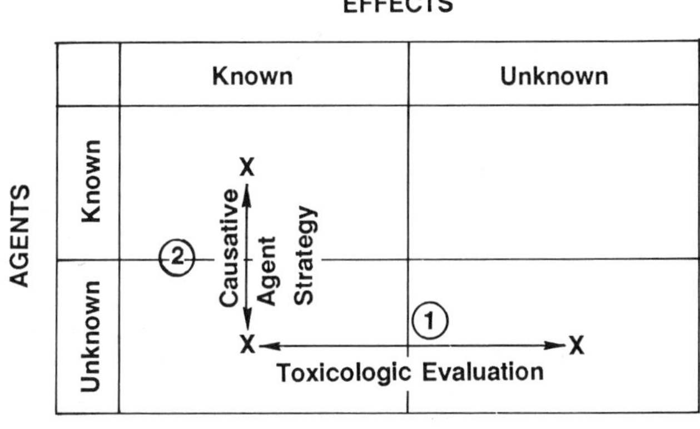

FIGURE 3-9 The selection of a strategy for toxicologic studies to increase knowledge of both agents and effects.

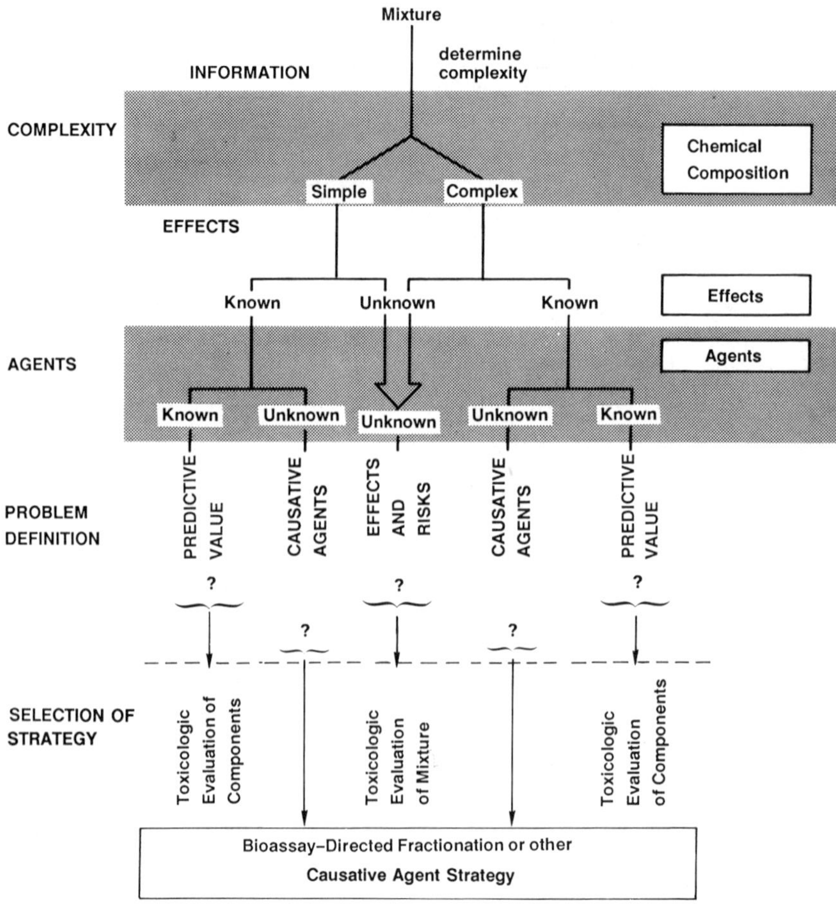

FIGURE 3-10 Illustration of effect of knowledge on selection of testing strategy and logical sequence from first determining effects of mixture and then determining causative or toxic agents.

REFERENCES

Alarie, Y. C., and R. C. Anderson. 1979. Toxicologic and acute lethal hazard evaluation of thermal decomposition products of synthetic and natural polymers. Toxicol. Appl. Pharmacol. 51:341–362.

Albert, R. E., J. Lewtas, S. Nesnow, T. W. Thorslund, and E. Anderson. 1983. Comparative potency method for cancer risk assessment application to diesel particulate emissions. Risk Anal. 3:101–117.

Andon, B., M. Jackson, V. Houk, and L. Claxton. 1984. Evaluation of Chemical and Biological Methods for the Identification of Mutagenic and Cytotoxic Hazardous Waste Samples. Health Effects Research Lab., Research Triangle Park, N.C. (Available from NTIS as PB84-201615.) (27 pp.)

Beck, L. S., D. I. Hepler, and K. L. Hansen. 1982. The acute toxicology of selected petroleum hydrocarbons, pp. 1–12. In H. N. MacFarland, C. E. Holdsworth, J. A. MacGregor, R. W. Call, and M. L. Kane (eds.). The Toxicology of Petroleum Hydrocarbons. American Petroleum Institute, Washington, D.C.

Bingham, E., R. P. Trosset, and D. Warshawsky. 1980. Carcinogenic potential of petroleum hydrocarbons. A critical review of the literature. J. Environ. Pathol. Toxicol. 3:483–563.

Bock, F. G., A. P. Swain, and R. L. Stedman. 1969. Bioassays of major fractions of cigarette smoke condensate by an accelerated technic. Cancer Res. 29:584–587.

Bos, R. P., J. L. G. Theuws, C. M. Leijdekkers, and P.T. Henderson. 1984. The presence of the mutagenic polycyclic aromatic hydrocarbons benzo[a]pyrene and benz[a]anthracene in creosote P1. Mutat. Res. 130:153–158.

Bridbord, K., and J. G. French. 1978. Carcinogenic and mutagenic risks associated with fossil fuels, pp. 451–463. In P. W. Jones and R. I. Freudenthal (eds.). Carcinogenesis. Vol. 3: Polynuclear Aromatic Hydrocarbons. Raven Press, New York.

Brown, E. M., and W. R. Hewitt. 1984. Dose-response relationships in ketone-induced potentiation of chloroform hepato- and nephrotoxicity. Toxicol. Appl. Pharmacol. 76:437–453.

Buhler, D. R., and M. E. Rasmusson. 1968. The oxidation of drugs by fishes. Comp. Biochem. Physiol. 25:223–239.

Carpenter, C. P., E. R. Kinkead, D. L. Geary, Jr., L. J. Sullivan, and J. M. King. 1975a. Petroleum hydrocarbon toxicity studies. I. Methodology. Toxicol. Appl. Pharmacol. 32:246–262.

Carpenter, C. P., E. R. Kinkead, D. L. Geary, Jr., L. J. Sullivan, and J. M. King. 1975b. Petroleum hydrocarbon toxicity studies. II. Animal and human response to vapors of varnish markers' and painters' naphtha. Toxicol. Appl. Pharmacol. 32:263–281.

Carpenter, C. P., E. R. Kinkead, D. L. Geary, Jr., L. J. Sullivan, and J. M. King. 1975c. Petroleum hydrocarbon toxicity studies. III. Animal and human response to vapors of Stoddard Solvent. Toxicol. Appl. Pharmacol. 32:282–297.

Carpenter, C. P., E. R. Kinkead, D. L. Geary, Jr., L. J. Sullivan, and J. M. King. 1975d. Petroleum hydrocarbon toxicity studies. IV. Animal and human response to vapors of rubber solvent. Toxicol. Appl. Pharmacol. 33:526–542.

Carpenter, C. P., E. R. Kinkead, D. L. Geary, Jr., L. J. Sullivan, and J. M. King. 1975e. Petroleum hydrocarbon toxicity studies. V. Animal and human response to vapors of mixed xylenes. Toxicol. Appl. Pharmacol. 33:543–558.

Carpenter, C. P., E. R. Kinkead, D. L. Geary, Jr., L. J. Sullivan, and J. M. King. 1975f. Petroleum hydrocarbon toxicity studies. VI. Animal and human responses to vapors of "60 Solvent." Toxicol. Appl. Pharmacol. 34:374–394.

Carpenter, C. P., E. R. Kinkead, D. L. Geary, Jr., L. J. Sullivan, and J. M. King. 1975g. Petroleum hydrocarbon toxicity studies. VII. Animal and human response to vapors of "70 Solvent." Toxicol. Appl. Pharmacol. 34:395–412.

Carpenter, C. P., E. R. Kinkead, D. L. Geary, Jr., L. J. Sullivan, and J. M. King. 1975h. Petroleum hydrocarbon toxicity studies. VIII. Animal and human response to vapors of "140° Flash Aliphatic Solvent." Toxicol. Appl. Pharmacol. 34:413–429.

Carpenter, C. P., E. R. Kinkead, D. L. Geary, Jr., R. C. Myers, D. J. Nachreiner, L. J. Sullivan, and J. M. King. 1976a. Petroleum hydrocarbon toxicity studies. IX. Animal and human response to vapors of "80 Thinner." Toxicol. Appl. Pharmacol. 36:409–425.

Carpenter, C. P., D. L. Geary, Jr., R. C. Myers, D. J. Nachreiner, L. J. Sullivan, and J. M. King. 1976b. Petroleum hydrocarbon toxicity studies. X. Animal and human response to vapors of "50 Thinner." Toxicol. Appl. Pharmacol. 36:427–442.

Carpenter, C. P., D. L. Geary, Jr., R. C. Myers, D. J. Nachreiner, L. J. Sullivan, and J. M. King. 1976c. Petroleum hydrocarbon toxicity studies. XI. Animal and human response to vapors of deodorized kerosene. Toxicol. Appl. Pharmacol. 36:443–456.

Carpenter, C. P., D. L. Geary, Jr., R. C. Myers, D. J. Nachreiner, L. J. Sullivan, and J. M. King. 1976d. Petroleum hydrocarbon toxicity studies. XII. Animal and human response to vapors of "40 Thinner." Toxicol. Appl. Pharmacol. 36:457–472.

Carpenter, C. P., D. L. Geary, Jr., R. C. Myers, D. J. Nachreiner, L. J. Sullivan, and J. M. King. 1976e. Petroleum hydrocarbon toxicity studies. XIII. Animal and human response to vapors of toluene concentrate. Toxicol. Appl. Pharmacol. 36:473–490.

Carpenter, C. P., D. L. Geary, Jr., R. C. Myers, D. J. Nachreiner, L. J. Sullivan, and J. M. King. 1977a. Petroleum hydrocarbon toxicity studies. XIV. Animal and human response to vapors of "High Aromatic Solvent." Toxicol. Appl. Pharmacol. 41:235–249.

Carpenter, C. P., D. L. Geary, Jr., R. C. Myers, D. J. Nachreiner, L. J. Sullivan, and J. M. King. 1977b. Petroleum hydrocarbon toxicity studies. XV. Animal response to vapors of "High Naphthenic Solvent." Toxicol. Appl. Pharmacol. 41:251–260.

Carpenter, C. P., D. L. Geary, Jr., R. C. Myers, D. J. Nachreiner, L. J. Sullivan, and J. M. King. 1977c. Petroleum hydrocarbon toxicity studies. XVI. Animal response to vapors of "Naphthenic Aromatic Solvent." Toxicol. Appl. Pharmacol. 41:261–270.

Carpenter, C. P., D. L. Geary, Jr., R. C. Myers, D. J. Nachreiner, L. J. Sullivan, and J. M. King. 1978. Petroleum hydrocarbon toxicity studies. XVII. Animal response to n-nonane vapor. Toxicol. Appl. Pharmacol. 44:53–61.

Clausing, P., and R. Bieleke. 1980. Aspects of methodology employed in the investigation of combined chemical effects on acute oral toxicity. Arch. Toxicol. Suppl. 4:394–395.

Crane, C. R., D. C. Sanders, B. R. Endecott, J. K. Abbott, and P. W. Smith. 1977. Inhalation Toxicology: I. Design of a Small Animal Test System. II. Determination of the Relative Toxic Hazards of 75 Aircraft Cabin Materials. FAA-AM-77-9. U.S. Federal Aviation Administration, Oklahoma City, Okla.

DeMarini, D. M. 1983. Genotoxicity of tobacco smoke and tobacco smoke condensate. Mutat. Res. 114:59–89.

Dichter, C. R. 1984. Risk estimates of liver cancer due to aflatoxin exposure from peanuts and peanut products. Food Chem. Toxicol. 22:431–437.

Feder, P. I., E. Margosches, and J. Bailar. 1984. A Strategy for Evaluating the Toxicity of Chemical Mixtures. Draft Report. EPA Contract No. 68-01-6721, Task 76. U.S. Environmental Protection Agency, Washington, D.C.

Florin, I., L. Rutberg, M. Curvall, and C. R. Enzell. 1980. Screening of tobacco smoke constituents for mutagenicity using the Ames test. Toxicology 15:219–232.

Heussner, J. C., J. B. Ward, Jr., and M. S. Legator. 1985. Genetic monitoring of aluminum workers exposed to coal tar pitch volatiles. Mutat. Res. 155:143–156.

Hewitt, W. R., and G. L. Plaa. 1983. Dose-dependent modification of 1,1-dichloroethylene toxicity by acetone. Toxicol. Lett. 16:145–152.

Hilado, C. J. 1976. Relative toxicity of pyrolysis products of some foams and fabrics. J. Combust. Toxicol. 3:32–60.

Hite, M., and H. Skeggs. 1979. Mutagenic evaluation of nitroparaffins in the Salmonella typhimurium/mammalian—microsome test and the micronucleus test. Environ. Mutagen. 1:383–389.

Hoffmann, D., and E. L. Wynder. 1971. A study of tobacco carcinogenesis. XI. Tumor initiators, tumor accelerators, and tumor promoting activity of condensate fractions. Cancer 27:848–864.

Houk, V. S., and L. C. Claxton. 1986. Screening complex hazardous wastes for mutagenic activity using a modified version of the thin-layer chromatography Salmonella assay. Mutat. Res. 169: 81–92.

Huff, J. 1982. Carcinogenesis bioassay results from the National Toxicology Program. Environ. Health Perspect. 45:185–198.

International Agency for Research on Cancer (IARC). 1986. IARC Monographs on the Evaluation of Carcinogenic Risk of Chemicals to Humans. Vol. 38: Tobacco Smoking. International Agency for Research on Cancer, Lyon, France.

Ishikawa, T., T. Shimamine, and S. Takayama. 1975. Histologic and electron microscopy observations on diethylnitrosamine-induced hepatomas in small aquarium fish (Oryzias latipes). J. Natl. Cancer Inst. 55:909–916.

Ishikawa, T., S. Takayama, and T. Kitagawa. 1978. Autoradiographic demonstration of DNA repair synthesis in ganglion cells of aquarium fish at various ages *in vivo*. Virchows Arch. B. 28:235–242.

Klaasen, C. D. 1986. Principles of toxicology, pp. 11–32. In C. D. Klaasen, M. O. Amdur, and J. Doull (eds.). Casarett and Doull's Toxicology: The Basic Science of Poisons. 3rd ed. Macmillan, New York.

Klimisch, H. J., H. W. M. Hollander, and J. Thyssen. 1980. Comparative measurements of the toxicity to laboratory animals of products of thermal decomposition generated by the method of DIN 53 436. J. Combust. Toxicol. 7:209–230.

Klotz, A. V., J. J. Stegeman, and C. Walsh. 1984. Multiple isozymes of hepatic cytochrome P-450 from the marine teleost fish scup (Stenotomus chrysops). Mar. Environ. Res. 14:402–404.

Lech, J. J., M. J. Vodicnik, and C. R. Elcombe. 1982. Induction of monooxygenase activity in fish, pp. 107–148. In L. J. Weber (ed.). Aquatic Toxicology. Vol. 1. Raven Press, New York.

Levin, B. C. 1986. A Summary of the NBS Literature Reviews on the Chemical Nature and Toxicity of the Pyrolysis and Combustion Products from Seven Plastics: Acrylonitrile-Butadiene-Styrenes (ABS), Nylons, Polyesters, Polyethylenes, Polystyrenes, Poly(Vinyl Chlorides) and Rigid Polyurethane Foams. NBSIR 85-3267. National Bureau of Standards, Gaithersburg, Md.

Levin, B. C., A. J. Fowell, M. M. Birky, M. Paabo, A. Stolte, and D. Malek. 1982. Further Development of a Test Method for the Assessment of the Acute Inhalation Toxicity of Combustion Products. NBSIR 82-2532. National Bureau of Standards, Gaithersburg, Md.

Levin, B. C., M. Paabo, J. L. Gurman, and S. E. Harris. In press. Effects of exposure to single or multiple combinations of the predominant toxic gases and low oxygen atmospheres produced in fires. Fundam. Appl. Toxicol.

Lewtas, J. 1983. Evaluation of the mutagenicity and carcinogenicity of motor vehicle emissions in short-term bioassays. Environ. Health Perspect. 47:141–152.

Lewtas, J., S. Nesnow, and R. E. Albert. 1983. A comparative potency method for cancer risk assessment: Clarification of the rationale, theoretical basis, and application to diesel particulate emissions. Risk Anal. 3:133–137.

Linsell, C. A. 1979. Decision on the control of a dietary carcinogen—aflatoxin, pp. 111–112. In W. Davis and D. Rosenfeld (eds.). Carcinogenic Risks: Strategies for Intervention. (IARC Scientific Publications No. 25.) IARC, Lyon, France.

MacDonald, J. R., A. J. Gandolfi, and I. G. Sipes. 1982. Acetone potentiation of 1,1,2-trichloroethane hepatotoxicity. Toxicol. Lett. 13:57–69.

MacFarland, H. N., C. E. Holdsworth, J. A. MacGregor, R. W. Call, and M. L. Kane, eds. 1984. Applied Toxicology of Petroleum Hydrocarbons. Advances in Modern Environmental Toxicology. Vol. VI. Princeton Scientific Publishers, Princeton, N.J.

Mehlman, M. A., C. P. Hemstreet III, J. J. Thorpe, and N. K. Weaver, eds. 1984. Renal Effects of Petroleum Hydrocarbons. Advances in Modern Environmental Toxicology. Vol. VII. Princeton Scientific Publishers, Princeton, N.J.

Milby, T. H., T. L. Miller, and T. L. Forrester. 1985. PCB-containing transformer fires: Decontamination guidelines based on health considerations. J. Occup. Med. 27:351–356.

Mitchell, C. L. 1966. Effect of morphine and chlorpromazine alone and in combination on the reaction to noxious stimuli. Arch. Int. Pharmacodyn. Ther. 163:387–392.

Mitchell, C. L. 1976. The design and analysis of experiments for the assessment of drug interactions. Ann. N.Y. Acad. Sci. 281:118–135.

NCI (National Cancer Institute). 1976. Report of the Subtask Group on Carcinogen Testing to the Interagency Collaborative Group on Environmental Carcinogenesis. NCI, Bethesda, Md.

Nesnow, S., L. L. Triplett, and T. J. Slaga. 1982a. Comparative tumor-initiating activity of complex mixtures from environmental particulate emissions on SENCAR mouse skin. JNCI 68:829–834.

Nesnow, S., C. Evans, A. Stead, J. Creason, T. J. Slaga, and L. L. Triplett. 1982b. Skin carcinogenesis studies of emission extracts, pp. 295–320. In J. Lewtas (ed.). Toxicological Effects of Emissions from Diesel Engines. Elsevier, Amsterdam.

Nishioka, M. G., C. C. Chuang, B. A. Petersen, A. Austin, and J. Lewtas. 1985. Development and quantitative evaluation of a compound class fractionation scheme for bioassay-directed characterization of ambient air particulate matter. Environ. Int. 11:137–146.

NRC (National Research Council). 1977. Principles and Procedures for Evaluating the Toxicity of Household Substances. National Academy of Sciences, Washington, D.C. (130 pp.)

NRC. 1983. Feasibility of Assessment of Health Risks from Vapor-Phase Organic Chemicals in Gasoline and Diesel Exhaust. National Academy Press, Washington, D.C.

NRC. 1984. Toxicity Testing. National Academy Press, Washington, D.C.

Pederson, T. C., and J. S. Siak. 1981. The role of nitroaromatic compounds in the direct-acting mutagenicity of diesel particle extracts. J. Appl. Toxicol. 1(2):54–60.

Petajan, J. H., K. J. Voorhees, S. C. Packham, R. C. Baldwin, I. N. Einhorn, M. L. Grunnet, B. G. Dinger, and M. M. Birky. 1975. Extreme toxicity from combustion products of a fire-retarded polyurethane foam. Science 187:742–744.

Plaa, G. L., and W. R. Hewitt. 1982a. Methodological approaches for interaction studies: Potentiation of haloalkane-induced hepatotoxicity, pp. 67–96. In H. F. Stich, H. W. Leung, and J. R. Roberts (eds.). Workshop on the Combined Effects of Xenobiotics, Ottawa, Ontario, Canada, June 22–23, 1981. NRCC No. 18978. National Research Council Canada, Associate Committee on Scientific Criteria for Environmental Quality, Ottawa, Canada.

Plaa, G. L., and W. R. Hewitt. 1982b. Potentiation of liver and kidney injury by ketones and ketogenic substances, pp. 65–75. In H. Yoshida, Y. Hagihara, and S. Ebashi (eds.). Advances in Pharmacology and Therapeutics II. Proceedings of the 8th International Congress of Pharmacology, Tokyo, 1981. Vol. 5: Toxicology and Experimental Models. Pergamon, New York.

Rosenkranz, H. S., E. C. McCoy, D. R. Sanders, M. Butler, D. K. Kiriazides, and R. Mermelstein. 1980. Nitropyrenes: Isolation, identification, and reduction of mutagenic impurities in carbon black and toners. Science 209:1039–1043.

Salmeen, I. T., A. M. Pero, R. Zator, D. Schuetzle, and T. L. Riley. 1984. Ames assay chromatograms and the identification of mutagens in diesel particle extracts. Environ. Sci. Technol. 18:375–382.

Scala, R. A. In press. Motor gasoline toxicity. Fundam. Appl. Toxicol.

Stanton, M. F. 1965. Diethylnitrosamine-induced hepatic degeneration and neoplasia in the aquarium fish, *Brachydanio rerio*. J. Nat. Cancer Inst. 34:117–130.

Stegeman, J. J., P. J. Kloepper-Sams, and J. W. Farrington. 1986. Monooxygenase induction and chlorobiphenyls in the deep-sea fish *Coryphaenoides armatus*. Science 231:1287–1289.

Williams, D. E., and D. R. Buhler. 1983. Multiple forms of cytochrome P-448 and P-450 purified from β-naphthoflavone fed rainbow trout. Abstract No. 3618. Fed. Proc. 42:910.

Witschi, H. P. 1982. Altered tissue reactivity and interactions between chemicals, pp. 263–288. In Assessment of Multichemical Contamination: Proceedings of an International Workshop, Milan, Italy, April 28–30, 1981. National Academy Press, Washington, D.C.

Yusa, S. 1985. Development of laboratory test apparatus for evaluation of toxicity of combustion products of materials in fire, pp. 471–487. In Seventh Joint Panel Meeting of the UJNR Panel on Fire Research and Safety. Proceedings. NBSIR 85-3118. National Bureau of Standards, Gaithersburg, Md.

4

Sampling and Chemical Characterization

This chapter presents a set of general guidelines and strategies to consider when designing a scheme for sampling and analyzing complex mixtures. No protocol or group of protocols for sampling or analysis will be applicable or appropriate for all types of mixtures. What follows is not intended to constitute a handbook of sampling and analytic methods for complex mixtures. Rather, researchers should design or modify methods as appropriate in each particular instance.

However, some general rules can be followed. All of them are based on the approach in which sampling of a complex mixture has three components: the relevance of the material collected to the human situation, the use to which the material will be put, and the potential for human exposure, including routes and extent of exposure and bioavailability.

CLASSIFICATION

It is important to know as much as possible about the chemical composition of a mixture before any useful toxicologic prediction can be undertaken (toxicologic testing can be performed on a mixture in the absence of chemical knowledge if prediction is not the objective). Information on the chemical and physical properties of a mixture can have a direct impact on the selection of testing strategies. Once the mixture has been partially defined, sampling techniques, bioavailability, and chemical characterization can be considered.

In light of the understanding that complex mixtures can vary widely—from combustion and distillation products of fossil or synthetic fuels, to pyrolysis products of tobacco or synthetic materials in buildings, to components leaking from waste-dump sites—some criteria must be identified for the mixture, regardless of origin. The first criterion, *physical state,* must be determined. One can encounter such complexity as a mixed-state situation, a slurry of particulate matter with a liquid or a gas or vapors and gases, which differs substantially from each fraction and is not toxicologically comparable with the "mix." The second, *chemical-class diversity and component multiplicity*, will have to be evaluated; this is difficult, in that a continuum of complexity can be encountered, and whether the components are known or unknown will influence the predictability of the composition and the biohazard. The third criterion, *stability of the mixture*, will be important, because chemical instability (i.e., the potential for chemical interactions) will seriously affect the reliability of the sample and the reproducibility of the toxicologic experiment to evaluate the mixture.

Complex mixtures can be divided according to whether they are derived from combustion or distillation products or from noncombusted materials. Examples of the first case are fossil and synthetic fuels and vegetable and synthetic materials. Examples of the second are food, water, and drugs; hazardous waste; municipal and sewage waste. If one knows the origin of a sample, one can deduce qualitative and quantitative information about the mixture in question, even if physical characterization and chemical characterization are minimal. Extensive information on chemical characterization of a series of fossil-fuel-related materials has been published (e.g., Wright and Dauble, 1986). That information makes it possible to predict chemical classes to be found in mixtures derived from these sources. A detailed discussion of the various origins of complex mixtures is found in Appendix A.

SAMPLING

Consideration of bioavailability (relevant to human exposure) and of ultimate sample use must be incorporated into the sampling strategy, if it is to be effective in producing materials for assay that will provide data relevant to human health. It is important also to consider the preservation of sample integrity; sample alterations must be minimized to ensure that the substance assayed is the substance presented to the human environment.

COLLECTION STRATEGY

Exposure

In designing a sampling protocol, input from several kinds of specialists is necessary. Toxicologists, industrial hygienists, or other experts in human health effects should be consulted regarding the most likely routes of exposure to the mixture in question. Exposure of humans via inhalation or ingestion of water or food is most common. In occupational settings, absorption through the skin can also be important. Exposure to a mixture via a combination of routes is not unusual. For example, particles might be inhaled and later swallowed after being cleared from the lungs by mucociliary action. It is usually practical to consider only the presumed primary route of exposure in designing a sampling protocol.

Meteorology

Seasonal or meteorologic considerations are particularly germane to the design of protocols for sampling environmental mixtures. Increases in air temperature can result in loss of benzo[a]pyrene (BaP), a carcinogenic polycyclic aromatic hydrocarbon (PAH), from particles collected during high-volume atmospheric sampling (De Wiest and Rondia, 1976). Seasonal variation can also affect surface areas and density of atmospheric particles (Corn et al., 1971; Flessel et al., 1984). Wastewater, soil, and waste-dump sample composition can be expected to vary seasonally. Changes in temperature, freezing conditions, and rainfall can all modify the water table and influence environmental sample composition.

Temporal Factors

Industrial hygienists and process engineers should be consulted as to temporal considerations of sampling. For industrial samples, the time course of production is important. One must consider whether to sample only at the end of a particular batch process, at the middle of a run, or throughout the production of the material in question. For example, it has been shown that the composition of coal-liquefaction recycle oils changes markedly during the first few days after startup of a process, but is relatively stable after that (Burke et al., 1984). In the case of air sampling at a manufacturing site, both the time of day and the point in the workweek are important. The variability of samples during a workweek has been illustrated in the monitoring of oncology nurses handling cytostatic drugs; the mutagenicity of material extracted from their urine increased toward the end of the week and decreased during the weekend (Nguyen et al., 1982).

The duration of sampling is important, particularly for emission materials. If the substances of interest are in low concentration, the sampling period must be long enough to collect material required for the assays proposed. This statement of the obvious must be balanced by two other considerations. First, it can take some preliminary investigation to ascertain the likely concentration and to determine whether the concentration of the material is relatively stable or likely to fluctuate. Second, the nature of the collected material might change as the capacity of the collection device is approached. Gorse et al. (1982) showed that the biologic activity of diesel exhaust particles changed as the collecting filter became loaded: the percentage of extractable material, the fluorescence of high-pressure liquid chromatography (HPLC) fractions, and the mutagenicity with respect to *Salmonella typhimurium* all increased linearly, presumably because the loaded filter became a more efficient collector of mutagenic vapors.

Spatial Considerations

Inclusion of spatial considerations in the sampling design requires the advice of experts, including industrial hygienists, engineers, and biostatisticians. The primary consideration should be the relevance of the sample site to potential human exposure. If the material is from a manufacturing or refining process, the design should ignore materials in closed systems in favor of open points of maintenance, ingredient addition, product removal, and so forth. Waste sites should be studied as to the availability of deposited materials. Personnel access to the site and the potential for materials to enter water supplies, soil, and food chains are important. In the case of emission, the nature of the research or regulatory questions asked will define whether sampling should focus on the point of discharge, some distant site, or both.

A testing program will probably require consideration of multiple testing sites for comparative studies. Whether sampling is to be purposive or of a random statistical design will be determined largely by the nature of the potential for human exposure.

Purposive sampling requires selection of a set of sites for generating samples of a particular type. In a study of coal-liquefaction biohazard potential, sampling sites would be at various process points: the coal-slurrying step, the reaction vessel, points of refining and separation, discharge of wastewater, final products, recycle materials, and waste. Similarly, sampling could be done in a waste-dump site with a history of material escape or near an emission source and at designated distances from it.

Random statistical designs seek to provide an overall picture of an area by sampling at a statistically determined number of places within that area. In a series of publications prepared for the Department of Energy, Gilbert (1983, 1984) described a number of sampling designs, generally termed "probability sampling" (see Table 4-1). In simple random sampling (see Figure 4-1), every unit of a target sample population (or geographic area) has an equal probability of being collected for assay. The units are numbered from 1 to N, and samples are chosen by drawing n numbers ($n < N$) from a random-number table or by use of a computer. The units designated by the selected numbers are sampled. This mode of sampling is best applied to a homogeneous population with no dominant spatial or temporal trends. Gilbert noted that most statistical inferential procedures assume that data were collected through simple random sampling. Sites of likely sampling of complex mixtures might not be homogeneous with regard to composition or bioavailability of components (e.g., waste

TABLE 4-1 Summary of Sampling Designs and When They Are Most Useful

Sampling Design	Most Useful When
Haphazard sampling	Population homogeneous over time and space essential; method not recommended, because of difficulty in verifying assumption of homogeneity
Judgment sampling	Target population well defined and homogeneous, so sample-selection bias is not a problem; or specific environmental samples selected for unique value and interest, rather than for making inferences to wider population
Probability sampling	
Simple random sampling	Homogeneous, i.e., no dominant trends or patterns
Stratified random sampling	Homogeneous within strata (subregions); might want to consider strata as domains of study
Systematic sampling	Trends over time or space must be quantified or strictly random methods are impractical
Multistage sampling	Target population large and homogeneous; simple random sampling used to select contiguous groups of population units
Cluster sampling	Population units cluster (schools of fish, clumps of plants, etc.); ideally, cluster means are similar in value, but concentrations within clusters should vary widely
Double sampling	Must be strong linear relation between variable of interest and less expensive or more easily measured variable

dumps). In addition, practical considerations in the collection of field samples can limit the use of simple random samples.

Another type of random sampling is stratified sampling. This assumes a number of nonoverlapping populations that differ from each other, but are internally homogeneous. The population or study site is divided into as many strata as appropriate, and simple random sampling is applied to each stratum. Another technique is multistage subsampling, which involves division of the site or population into primary units (Cochran, 1977). A set of primary units is chosen by simple random sampling, and samples are collected within each of these, under the direction of simple random sampling. This type of design is particularly applicable to geographic areas with variations in terrain, habitat, or other factors that require comparison. It is also useful for a relatively homogeneous, but large, study site, if the number of samples that can be collected is limited. In cluster sampling, which is related, population units are grouped into clusters, a number of clusters are randomly selected, and all units within the selected clusters are sampled.

Distinct from the random designs are the methods of systematic sampling. Gilbert (1984) indicated that these methods often find use in environmental monitoring, because they are generally easier to implement under field conditions. In addition, data from statistical investigations have indicated that sys-

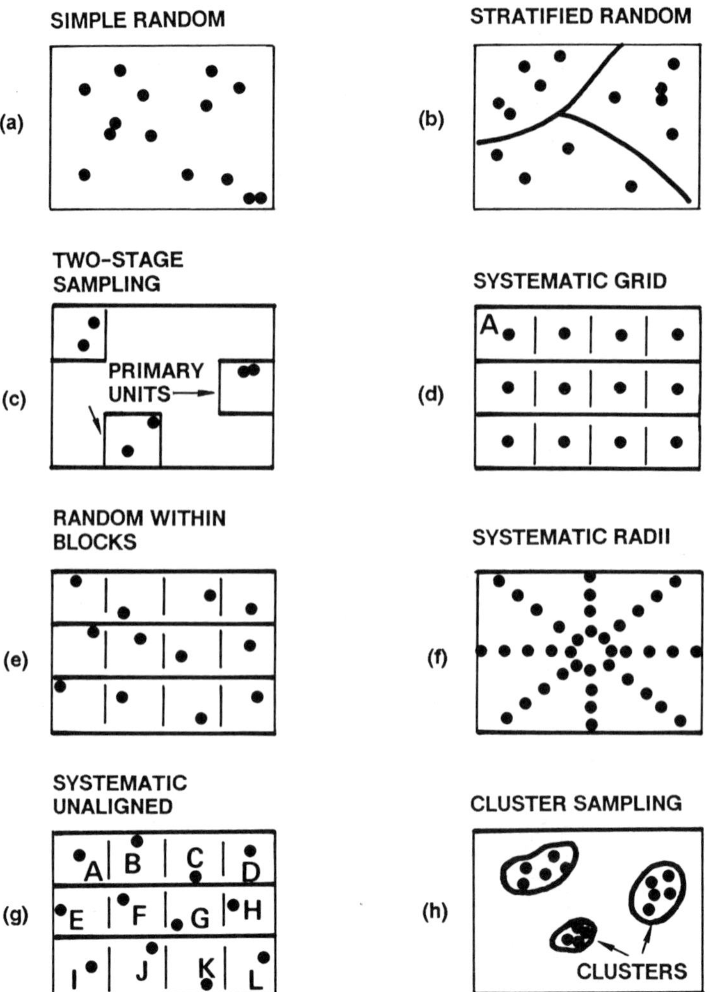

FIGURE 4-1 Some probability field sampling designs. Reprinted with permission from Gilbert (1983).

tematic sampling can be superior to simple or stratified random sampling for some types of environmental variables. In general, systematic sampling begins with random selection of one unit of the study population. This serves as the starting point of a sampling pattern according to some established spatial or temporal frequency. A problem common to systematic designs is that, if the variable being measured is subject to periodicity or cycles, misleading results will be obtained. Gilbert cited another difficulty: the accuracy of estimates of sampling errors and other statistics depends on the study population's being random.

In the aligned-square grid design, a location (or other unit) is chosen at random. The location of each sample is then chosen by applying a grid of fixed dimensions over the chosen area, and two random coordinate numbers are drawn to fix the location of the original point. Each sample site is then fixed in relation to this original point. Variations include the use of a triangular grid and the unaligned-grid-pattern design. In the latter (Figure 4-2), a point A is chosen randomly, and *X* and *Y* coordinates are established. To set points B, C, and D, one uses the *X* coordinate of A and three new randomly chosen *Y* coordinates. To set points E through I, the point A coordinate *Y* and random *X* coordinates are used.

One can also design systematic sampling methods based on lines. Figure 4-1 illustrates periodic sampling of radii from a central point, such as a known discharge source.

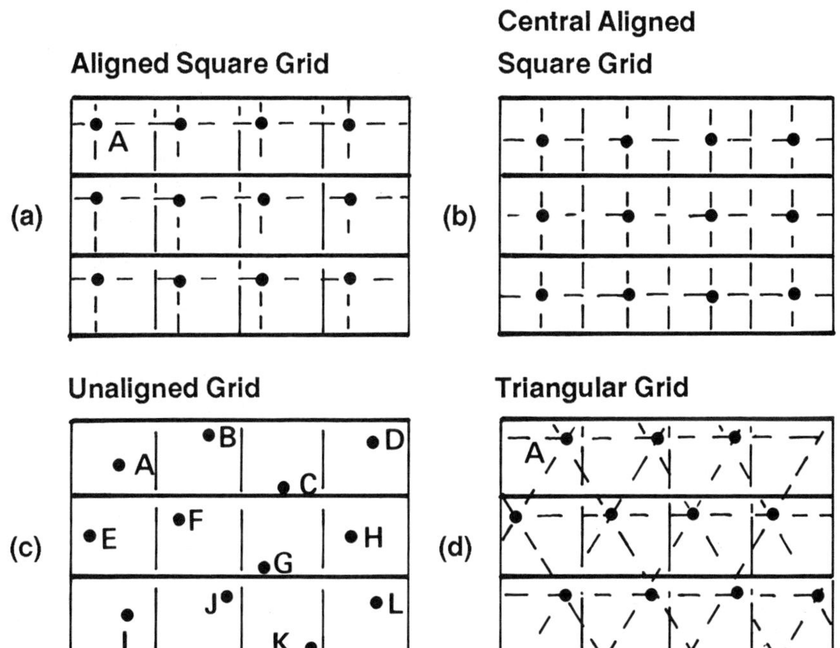

FIGURE 4-2 Some systematic designs for sampling in space. Reprinted with permission from Gilbert (1984).

Collection Procedure

Designing the collection process requires the input of a toxicologist to provide information on the nature of the assays and of a chemist and an engineer to produce a protocol that will efficiently provide a relevant sample material. One type of information to be supplied by a toxicologist is the amount of material that needs to be collected. Testing protocols geared to chemical analysis generally require smaller samples than more biologically oriented approaches. In vitro tests are more sparing of sample than are whole-animal studies.

The collection procedure depends largely on the source of the material and its physical state. Environmental samples are likely to be mixtures of materials in various physical states. They might include mixtures of gases, aerosolized liquids, and suspended particles; combinations of liquid and solid waste; biologic tissues, which are largely water; and so forth. We present here examples of sample types and methods deemed appropriate for their collection. Many of these were described at greater length in a recent EPA document (U.S. EPA, 1985), and some useful techniques were also reviewed by Alfheim et al. (1984).

Gases and Aerosols

Vapor-phase organic chemicals have been collected from ambient air, combustion exhaust gases, cigarette smoke, and indoor spaces for chemical characterization and bioassay (Hanson et al., 1981, 1984; Griest et al., 1982; Higgins et al., 1983; Pellizzari et al., 1976; Pellizzari, 1982; Hughes et al., 1980; Krost et al., 1982). Chemical transformations and perturbations of the original chemical mixture can occur during the sampling step (Berkley and Pellizzari, 1978; Pellizzari et al., 1984; Pellizzari and Krost, 1984).

Air particles are a common source of mixtures that have environmental or health significance. Alfheim et al. (1984) recommended that particle-size fractionation be included in the collection scheme to increase the biologic relevance of a particulate sample. The International Standards Organization (ISO, 1983) and the American Conference of Governmental Industrial Hygienists (Lippmann, 1985) have established cut sizes appropriate to fractionation. They depend on the toxicity at deposition sites in the airways or at sites along the clearance pathways or storage sites. Inspirable particles are those which can be aspirated by the nose or mouth. Thoracic particles are those which can enter the thorax. Respirable particles are those which penetrate the lung's conductive airways. Another reason for size separation is that the chemical composition and biologic activities are related to particle size.

Among urban air particles, small size is associated with the presence of acidic compounds, and larger sizes with basic groups (Miller et al., 1979). In comparative assays, most mutagenic activity has been associated with smaller particles, less than 2.5 μm in diameter (Talcott and Harger, 1980; Preidecker, 1980; Chrisp and Fisher, 1980). Most of the total PAH content of air particles is found in particles less than 5 μm in diameter. It has been recommended, however, that results of assays of particles collected with size fractionation be compared with results of assays of particles collected without fractionation. Artifacts of size separation reportedly have reduced activity in mutagenicity assays (Alfheim et al., 1983), whereas other chemical transformations reportedly have increased mutagenicity (Clark et al., 1981).

Point-source sampling for air particles includes use of impactors (Cheng and Yeh, 1979), cyclones, electrostatic precipitators, and filters (U.S. EPA, 1985). Sampling methods have been described by Brusick and Young (1982) and Lentzen et al. (1978). Collection of particles from ambient air generally uses high-volume samplers (U.S. EPA, 1971), so-called massive-volume samplers (Henry et al., 1978; Cheng et al., 1984), and medium-, low-, or ultrahigh-volume samplers (Fitz et al., 1983). It has been recommended (U.S. EPA, 1985) that preparation of particulate samples for bacterial-mutagenicity monitoring begin with collection of standard high-volume samples with inert filter-collection media. For collection of larger amounts of material or when circumstances dictate a shorter collection period, use of an ultrahigh-volume sampler would be appropriate (U.S. EPA, 1985).

If gaseous emission is collected at high temperature, there will be less opportunity for condensation of materials on particles. To collect a gaseous mixture that adequately reflects the material being produced or emitted to the human environment, the possible presence of volatile materials must be considered in the collection scheme. When it is feasible, a gas can be cooled before introduction to an adsorbent material. Alfheim et al. (1984) indicated that XAD-2 resin was most commonly used for this purpose. Gaseous samples can thus be reduced to collected particles, extracts from adsorbents, and condensates. The testing of volatile materials not captured that way is a subject for research and is not generally applicable to routine environmental monitoring or widely used in bioassays.

Aqueous Materials

Aqueous samples may be collected in a relatively homogeneous state or may contain suspended solids. The most common method of sampling water is manual-grab collection of the required volume. Water generally has low concentrations of biologically active materials, which must be concentrated for application in most tests. Concentration protocols are therefore identified as the major part of the collection procedure. Methods for concentration of water with less than 5% associated solids, particularly drinking water, have been well described (Kopfler, 1980; Jolley, 1981; NRC, 1982). Aqueous solutions can be concentrated by removing the water via freeze concentration, lyophilization, vacuum evaporation, and reverse osmosis (Shapiro, 1961; Baker, 1970; Dawson and Mopper, 1978; Crathorne et al., 1979; Jolley et al., 1975; Kopfler et al., 1977). These methods concentrate organic and inorganic materials together, a procedure that Kopfler (1980) claimed was a disadvantage if the bioassay system to be used could not tolerate high concentrations of inorganic materials.

Another scheme involves selective concentration of organic materials from drinking water by adsorption of the contaminants on activated carbon or XAD resin columns (Middleton et al., 1956; Kopfler, 1980; McGuire and Suffet, 1983; U.S. EPA, 1985). A variety of solvents can be used for elution of organic substances from the activated-carbon columns. Supercritical liquid carbon dioxide is an efficient elution solvent (Modell et al., 1978). This method can be used for large-scale processing of water samples and results in the recovery of gram quantities of contaminants. The limitations of these techniques include the nonextraction or nonrecovery of inorganic materials and highly polar materials, ionic organic species, or volatile low-molecular-weight organic compounds. Methods that might be suitable for collection of these substances were cited by EPA (1985). Kopfler (1980) noted that organic acids and bases might be amenable to collection on resin columns after adjustment of the water pH to suppress ionization. Altering the pH of a mixture of materials, however, also results in changing its chemical composition or biologic activity.

Aqueous samples with large amounts of suspended solids are generally best prepared by separation of the phases. Gravity partitioning is done by storing the sample at 4°C for 24 hours. The solid phase can then be removed and processed separately for assay. Alternatively, high-pressure filtration or high-speed centrifugation can be used for phase separation. The liquid portion can be separated into aqueous and nonaqueous components, concentrated if necessary, and assayed. Some samples with suspended solids are amenable to processing by liquid-liquid extraction methods. Details of some of these protocols were described by EPA (1985).

Nonaqueous Materials

Many environmental and industry-related samples are nonaqueous liquids, including organic liquids, light and heavy oils, and some tars. Samples can contain mixtures of volatile (b.p., 36–100°C), moderately volatile (b.p., 100–300°C), and nonvolatile (b.p., over 300°C) materials, and there might be associated solids. Generally, the composition of such materials is poorly characterized and subject to variation after sampling. Much of the information on samples of this sort has been derived from research on fossil-fuel materials (Wright and Dauble, 1986). Sampling itself is usually not difficult; grab sampling often suffices. The samples are generally concentrated enough for assay of undiluted or unprocessed material. The preparatory issue with these materials therefore is not sampling, but rather treatment to make the sampled material compatible with the bioassay in question. Some of these issues are discussed later.

Solids and Sediments

Solid materials appear to be amenable to simple sampling techniques. However, the entire mass of material collected might not constitute the sample of interest. Soils and sediments are examples of materials wherein an unknown or small concentration of a substance of biologic interest is associated with a large quantity of matrix. The protocol for these materials must therefore include extraction and concentration. Specific recommendations for treatment of soils were made by EPA (1985).

Efforts should first be made to provide a homogeneous sample. Aggregates can be broken by crushing or cutting. To avoid a particle-size bias, the material can be quartered. This is done by spreading out the material on a clean surface and mechanically dividing the sample into four parts; three parts are returned to the storage vessel, and the remainder is quartered until the desired amount is obtained.

Sediments might contain water, which can be removed as described previously. The next step should be extraction in a Soxhlet apparatus with highly pure organic solvents. The nature of material to be recovered will depend largely on the solvents used. Solvents used for extracting materials from soil matrices for chemical analysis have included the following: benzene, ethyl acetate, and benzene-methanol-acetone (2:1:1) for humic substances (Ogner and Schnitzer, 1970; Cifrulak, 1969); *n*-pentane, carbon tetrachloride, and methylene chloride for oil (Jobson et al., 1974; Jensen, 1975; McGill and Rowell, 1980); and hexane-isopropanol (3:1) for polychlorinated biphenyls (Carey and Gowen, 1978).

Other extracting solvents have been used specifically to produce samples that would be amenable to biologic testing. Mutagenic materials were obtained from agricultural soils extracted with hexane-acetone (2:1) and dichloromethane (Göggelmann and Spitzauer, 1982; Brown et al., 1985). Likewise, mutagenic sediment extracts have been obtained through sequential extraction with diethyl ether and methanol (Kinae et al., 1981). Ether has also been used on sediments (Sato et al., 1983). The choice of solvents can likely be dictated solely by the nature of the material to be recovered, if the solvent can be evaporated or otherwise removed before bioassay. EPA (1985) has recommended rotary evaporation. The pH of the soil or sediment will be a major influence on the types of compounds extracted.

For most other types of solid samples, the primary preparative consideration will be to constitute a material compatible with the bioassay selected. The sample must be representative of the environmental or human exposure poten-

TABLE 4-2 Samplers Recommended for Various Types of Solid Waste[a]

Waste Type	Waste Location or Container	Sampling Devices
Slurry	Tank, bin, pit, pond, lagoon	Weighted bottle, dipper
Dry solid	Drum, sack, pile, truck, tank, pit, pond, lagoon	Thief, scoop, shovel
Sticky or moist solid and sludge	Drum, tank, truck, sack, pile, pit, pond, lagoon	Trier
Hard or packed waste	Drum, sack, truck	Auger

[a]Data from U.S. EPA (1982).

tial. This can be ensured by taking multiple samples (multiple sites on repeated occasions) and combining them. EPA (1985) has suggested that heating the material or conducting the sampling procedure at a high temperature might increase the homogeneity of some solids. Mixing is to be done before taking the sample, if that is feasible. EPA (1985) also cautioned that heating might increase volatilization of some sample components and alter characteristics of others. Table 4-2 lists types of sampling devices recommended for solid wastes.

Maintenance of Sample Integrity

When a sample has been collected, further safeguards must be in place if it is to be assumed that the data from the bioassay are appropriate for evaluating the sample's hazard potential. Mechanisms must be instituted at each step of sample handling to ensure that the material collected is the material tested. Balanced against this need to maintain sample integrity is the need to deliver the sample to the bioassay appropriately. The delivery must be done so that there is even distribution of the test item. It is also important that the test organisms not be killed and that they not be affected in any way that would impinge on the biologic end point to be measured. It is likely that all these conditions could be met only in an ideal world. Nonetheless, intelligent consideration of factors bearing on sample integrity is necessary.

Storage

A suitable means of sample storage is the first consideration. In all but the rarest of instances, there is a period between collection of the sample and its application in a biologic test. The researcher will want to preserve an aliquot for future reference, retesting, and other possible characterization. Storage conditions must be tailored to particular samples. Some general suggestions have been made for various classes of mixtures (U.S. EPA, 1985).

Factors that must be considered for storage conditions include the following:

• *Chemical transformations*. Potentially biohazardous substances are likely to be chemically reactive themselves or readily converted to reactive forms. In some instances (as with fossil-fuel mixtures), biologic activity has been shown to be lost with time after sampling (Schoeny, 1985). Temperature plays an important role; storage under liquid nitrogen or at −70°C enhances preservation.

• *Photochemical reactions*. Components of complex mixtures might be subject to photoactivation or photodegradation by near-UV light that would alter the biologic activity (Larson et al., 1977; Barnhart and Cox, 1980; Selby et al., 1983; Kalkwarf et al., 1984). It would be prudent to store samples in dark brown or actinic glass containers.

• *Microbial transformations*. Various species of bacteria and fungi can metabolize organic compounds found in water and soil samples (Cerniglia, 1981). Preservation might require autoclaving, although heat (dry or steam sterilization) can alter the contaminants. Liquid samples, such as water without suspended sediments, can be filter-sterilized.

• *Physical state*. It must be recognized that heterogeneous materials can be expected to separate into phases during storage. Particles will settle out of liquids; volatile materials can escape; and aqueous and organic phases might separate, as might organic phases from other organic phases.

Preparation for Assay

In preparing test materials for assay, it is of the greatest importance to consider their mode of administration to the bioassay. The sample of interest might require modification of physical state for compatibility with the bioassay design. The most widely applied modification is solubilization of the sample in a substance that can be safely used for its delivery. The choice of solvents depends on the chemical properties of the mixture. For less complex mixtures of similar substances, such solvents as dimethyl sulfoxide, ethanol, methanol, acetonitrile, or methylene chloride might be adequate for most chemical or toxicologic studies. More complex mixtures (coal tar or liquids derived from coal liquefaction) might require a mixed solvent or a combination of solvents with increasing polarity. The failure of these to solubilize the entire mixture might affect the overall results. An estimate of the nonsolubilized (and ultimately nontested) material should be included in the sample evaluation.

For all chemical and toxicologic studies, solvent controls are necessary to gauge the effects of the delivery system on the assay outcome or test-organism viability. The solvent must be of the highest grade practicable (e.g., spectroscopic- or pesticide-grade).

Sample preparation methods, like sample collection and maintenance, must be tailored to specific sample types. Solubilization and extraction procedures should be undertaken in glass containers to minimize contamination. Collected particles and gases can also be extracted and the adsorbed materials assayed independently of the matrix. One recommended method entails the extraction of 1.0 g of a sample twice with dichloromethane either by Soxhlet extraction or by sonication followed by filtration and concentration. When these types of extractions are conducted, solvent incompatibility between chemical fractionation and bioassay systems can occur and might require solvent exchange (Burton et al., 1979; Springer et al., 1982, 1984; Hackett et al., 1983; Kelman and Springer, 1981; Mahlum, 1983a,b; Mahlum and Springer, 1986).

Quality Assurance

Quality assurance quantifies and documents the extent to which measures taken to maintain sample integrity have been successful. Information on sample source, collection, separation, and analysis should be recorded in a coherent and consistent fashion in a language devoid of jargon to facilitate information interchange among the members of the research team. The precision of the analytic method is assessed to evaluate the variability in the technique and requires the assay of independent replicate samples collected in the same manner at the same place and time. EPA has recommended that, for large sample lots, a fixed frequency of replicate measurements (e.g., 1 in 10 or 1 in 20) be chosen. For small sample lots, a greater frequency of repetition will be necessary (U.S. EPA, 1984).

Quality assurance also extends to the preparation of a sample for chemical or toxicologic evaluation, because, in addition to the mixture's being complex, numerous steps in separation, extraction, and fractionation could give rise to variable results. The addition of a known surrogate or reference material to the original complex mixture, to act as an internal standard for later extraction and evaluation, is useful. Distinctive chemicals, having physical and chemical properties similar to those found in the complex mixture, should be introduced at the collection site, thereby allowing an assessment of sample change during handling, transportation, and storage, as well as recovery during sample preparation. However, these internal standards should be carefully selected, because they might interfere with the anticipated bioassay.

ANALYSIS

The following discussion on chemical characterization of complex mixtures presents highlights of past and current analytic strategies and techniques. It introduces analytic tools of the future that can play an important role in more comprehensive characterization of mixtures and identification of components. This section should be regarded as a conceptual guide to mixture classification.

Chemical Classification

To make toxicologic predictions, analytic chemists and toxicologists need to know which chemical classes are present in a mixture. Toxicologists would be interested in major chemical classes—such as PAHs, aromatic amines, nitrosamines, halogenated hydrocarbons, direct-acting alkylating agents, and metals—because they can indicate the types of causative agents that might be present in the mixture. Analytic chemists would be interested in the similarities between compounds within a chemical class, because they can provide clues as to appropriate analytic procedures and strategies that could be used for sample fractionation, component separation, and identification.

Organic compounds can be divided into aliphatic, aromatic, cyclic aliphatic, and organometallic compounds, and inorganic compounds can be divided into metallic and nonmetallic elements and compounds. (The presence of heteroatom-containing compounds in each class is also important.)

Aliphatic compounds can be further divided into compounds with one carbon atom—such as methane, methanol, methylamine, methylnitrosamine, and methylene chloride—and compounds with two or more carbon atoms. The reactivity, volatility, stability, and solubility of a mixture can depend heavily on the numbers of carbon atoms in its aliphatic constituents. Aromatic compounds can be divided into one-ring compounds—such as benzene, toluene, anisole, thiophene, pyrrole, and aniline—and compounds with two or more rings. Many cyclic aliphatic compounds are solvents, such as cyclopentane, cyclohexane, decalin, and pyrrolidine. In most schemata, these compounds are not included, although they are present in some complex mixtures. In organometallic compounds, a metal atom is bound directly to two or more carbon atoms (e.g., organotin). This chemical classification for the aliphatic and aromatic compounds, and so on, is important when considering the strategies to be used in the chemical and toxicologic analysis of the mixture.

Although this type of chemical classification limits the number of classes, irrespective of the degree of complexity of the mixture in question, it permits the chemical and toxicologic aspects of the research to be interactive, in that only one chemical nomenclature needs to be used.

General Considerations in Fractionation of Complex Mixtures

Many approaches have been used for analyzing complex mixtures. For developing data bases on toxicity, standardized strategies on separation and chemical analysis might be beneficial, especially if comparisons are anticipated (Guerin et al., 1983; Haugen and Stamoudis, 1986; Guerin, 1981; Wilson et al., 1981; Wright et al., 1985; Benson et al., 1984; Hanson et al.,

1980, 1982, 1985; Stamoudis et al., 1983; Later et al., 1981, 1983a,b; Smith, 1983; Weimer and Wilson, 1983).

Requirements for Analysis

As part of selecting a strategy for chemical fractionation and characterization of complex mixtures, the in vitro and in vivo bioassay requirements must be clearly delineated. Toxicologists and chemists should collaborate to address the needs of the testing strategies and logistics discussed here.

Whether the purpose is to identify the causative agents in a mixture (Claxton, 1982; Bridbord and French, 1978; Hite and Skeggs, 1979; Florin et al., 1980) or to predict toxicity, some fundamental information is needed in designing chemical characterization experiments. Consideration should be given first to the total quantity of material (mass) and the mode of sample administration (fractions and subfractions). This preliminary information is necessary to set up the analytic and preparative procedures to support toxicity testing. For example, the analyte concentrations in the extracts, the quantity of material needed for each assay, and the total amount of material to be administered should be calculated. Generally, much more material is needed for a bioassay than for chemical analysis.

The route of administration must be considered, because it defines the delivery system that can be used. Is the route of administration to be inhalation, ingestion, topical application, or injection? The route of exposure for in vivo tests should simulate the exposure of humans. Thus, the mode of administration can be inhalation for in vivo experiments, but exposure can be to either gases, aerosols, or particles for in vitro tests. The physical state of the mixture, including particle or droplet size for inhalation exposure, must be defined; it will have a bearing not only on the form of administration, but on the predictive power of the test. Engineering factors, particularly in inhalation experiments for chamber design, must be considered, so that the exposure will simulate environmentally and biologically important conditions. In general, extraction, fractionation, and subfractionation of mixtures for chemical analysis are conducted on a relatively small scale, compared with those for bioassays. An analyst must be aware of the bioassay and chemical-analysis requirements during the application of isolation, fractionation, and subfractionation schemes.

Analysts have a repertoire of techniques available for elucidating chemical structures and quantifying analytes in mixtures. They consider mass requirements for successful analysis, the physical state of materials to be analyzed, the mode required to introduce an analyte or mixture into an analytic system, and the degree of complexity in a mixture (e.g., how much purification is needed).

Some analytic techniques can resolve simple mixtures, such as pesticides, to the extent that qualitative and quantitative analysis can be performed on individual analytes. The extent to which this can be achieved with an analytic technique should be recognized, so that sufficient fractionation and subfractionation are conducted in advance.

Some spectroscopic techniques are ultrasensitive and have low mass requirements (e.g., emission spectroscopy), whereas other techniques need relatively large quantities (e.g., nuclear magnetic resonance). The selectivity and specificity of a spectroscopic technique also determine the extent of purification required before analysis.

All four factors—quantity, physical state, mode of introduction, and purification requirements—should be thoroughly examined before approaches are chosen for creating fractions and subfractions.

All the issues referred to here should be noted in concert with the bioassay requirements, to ensure a compatible bioassay and chemical analysis approach. For information from a bioassay to correlate with information from a chemical analysis, the mixtures, fractions, and subfractions must have a common denominator; that is, the same chemical separation procedures should be used for both bioassay and chemical analysis (Claxton, 1982; Later et al., 1981; Guerin et al., 1980, 1983; Ho et al., 1980; Pellizzari et al., 1978).

Separation and Preliminary Examination

The identification of analytes in a mixture involves separation into individual components and structural elucidation with spectroscopic and spectrometric techniques. Separation and purification of the constituents are necessary because it is rarely possible to identify the constituents of a mixture directly. Separation of the constituents of a mixture should be quantitative for both bioassay and chemical characterization. It is also important that the components be in as pure a state as possible to avoid erroneous structural assignments. Methods for separating and purifying components of a mixture should not chemically alter any components. As indicated elsewhere, the history or origin of a mixture (see Appendix A) can reveal important information that aids in selecting separation processes and, to some extent, spectroscopic or spectrometric techniques. Preliminary review of the origins of a chemical mixture, its sampling characteristics, and its sample history will help in selecting a strategy for chemical separation and analysis.

Extraction and Concentration

Solvents and extraction methods used for various complex mixtures have recently been reviewed by several investigators (Epler, 1980; Chrisp and Fisher, 1980). Liquid and gaseous samples can be chemically analyzed as is; sometimes a concentrated sample is required (Hughes et al., 1980; Higgins et al., 1984; Flotard, 1980). Liquid-liquid extraction is suitable when small sample volumes are needed (Alfheim et al., 1984; Boparai et al., 1983). Dichloromethane and diethyl ether are commonly used for complex mixtures in aqueous samples (Maruoka and Yamanaka, 1980; van Hoof and Verheyden, 1981; Kringstad et al., 1981). Impurities or preservatives in the solvent should be noted, lest they be concentrated with the sample components and cause inappropriate interpretations of sample components. For example, peroxides and cyclohexene are present in ether and dichloromethane, respectively. Diethyl ether extraction of aqueous samples has been extensively used. Many solvents (methanol, dichloromethane, acetone, and dimethyl sulfoxide) have been used to extract and elute adsorbed material from particulate matter and analytes adsorbed on sorbent materials (Kool et al., 1981a,b; Epler, 1980; Chrisp and Fisher, 1980). Mixtures of solvents or sequential solvents with increasing polarity have also been used to elute analytes. The matrix of the sample limits both the choice of solvent and the method of extraction for all mixtures.

SEPARATION

Two principal approaches—liquid-liquid partitioning and chromatography—have been used to separate complex mixtures into chemically distinct fractions that are suitable for chemical testing or toxicity testing (Alfheim et al., 1984). These approaches are also used in fractionation for bioassay. Conventional methods are listed in Table 4-3. Each method has a different principle of action. Fractionation of complex extracts might require a combination of these methods.

Two important characteristics of fractionation methods are the accuracy and reproducibility of analyte recovery. When the recovery of mass is poor, the

TABLE 4-3 Conventional Separation Methods

Type	Variations	Materials
Liquid-liquid partitioning	Solvent-solvent; solvent-acid/base	Organic solvents of different polarity; aqueous acid/base solutions
Chromatography	Column (open), thin-layer (TLC)	
	Adsorption	Aluminum, silica gel, polyamide, Florisil
	Partition	Cellulose
	Molecular exclusion	Porous polymers, gels
	Ion-exchange	Ion-exchange resins (anionic, cationic)
	High-performance liquid (HPLC)	
	Adsorption	Aluminum, silica gel, polyamide, Florisil
	Partition	Cellulose, bonded phases
	Molecular exclusion	Porous polymers, gels
	Ion-exchange	Ion-exchange resins (anionic, cationic)

method should obviously be abandoned. The recovery of toxic activity, however, is more complicated to interpret (Alfheim et al., 1984). There might be a lack of additivity in a fractionated complex mixture, because of the loss of toxic analytes, because chemical reactions transform toxic analytes to nontoxic substances (or vice versa), or because of the nature of the assay itself. The toxic activity of a reconstituted mixture (a sample made by combining the individual fractions) is often compared with that of the original complex mixture (Schoeny et al., 1986). The sum of the toxic activities of the individual fractions might reveal effects of interactions, such as additivity and antagonism, between substances (Alfheim et al., 1984). Thus, it is important that both reproducibility and precision be established for the fractionation scheme. Spillover of an analyte into various fractions must be minimal, so that toxicity-test results are not obscured.

Liquid-Liquid Partitioning

This method has been applied to particulate organic matter (Daisey et al., 1980; Pellizzari et al., 1978), fly ash (Chrisp and Fisher, 1980), coal gasification and liquefaction products (Schoeny et al., 1980), and water samples (Kopfler et al., 1977).

Acid-base fractionation has been applied to many different types of complex mixtures (Hughes et al., 1980; Pellizzari et al., 1978; Swain et al., 1969; Bock et al., 1969; Wynder and Wright, 1957; Kier et al., 1974). It has the following advantages:

- It generally yields a good separation of chemical classes.
- It yields knowledge of chemical classes expected in various fractions, because the principles of the method are well understood and can be correlated with the physical and chemical properties of compounds.
- It is applicable to small and large samples.
- It can be applied to a wide variety of sample matrices and to organic extracts, whether lipophilic or hydrophilic.

Liquid-liquid partitioning is very useful for making preliminary assessments of toxicity data (Alfheim et al., 1984). Because some chemical knowledge can be inferred from the fractions derived by liquid-liquid partitioning, toxic activity in a fraction can be correlated with general chemical-class functionalities (Ho et al., 1980; Later et al., 1981; Later and Lee, 1983; Toste et al., 1982; Wright et al., 1985).

The disadvantage of liquid-liquid partitioning is that the process does not clearly define the chemical classes in each solvent used; instead, spillover often occurs between the individual fractions (i.e., an analyte appears in more than one fraction). Several investigators have observed this problem (Moller and Alfheim, 1983; Teranishi et al., 1978). In addition, the analyst should be cognizant of potential chemical reactions (e.g., hydrolysis, dehydrohalogenation, elimination, and oxidation) when a complex mixture is in contact with acids and bases. Such reactions might alter the bioactive compounds present in the original sample or even create new toxic compounds. Reports on extracts of particles from various sources have indicated that such reactions can occur (Teranishi et al., 1978). Practical disadvantages are that the procedure is generally laborious and that material can easily be lost by adsorption or volatilization when solutions are transferred from one container to another.

Several liquid-liquid partitioning schemes have been used in combination with toxicity testing to avoid the use of acids and bases (Dehnen et al., 1977; Hoffmann et al., 1980; Guerin et al., 1978; Chan et al., 1981). This reduces the potential for many undesired chemical reactions.

Chromatographic Separation

The general benefits of chromatographic separation methods lie in the vast choice available for application in the presence of different chemical and physical properties. The methods should be chosen to lead to only minimal irreversible chemical reactions and analyte losses.

Chromatographic separations can be divided into adsorption, partitioning, molecular exclusion, and ion exchange, as shown in Table 4-3, for open-column chromatography (Browning, 1969). In adsorption, there is competition between solid and gas (gas-solid chromatography, GSC) or between solid and liquid (column, thin-layer). Partitioning involves competition between liquid and gas (gas-liquid chromatography, GLC) or between liquid and liquid (column, paper, thin-layer). Gel-permeation chromatography in its purest form is based on size exclusion; however, it rarely is a homogeneous process, and instead can be a mixture of size exclusion, adsorption, and partitioning. Modern forms of chromatography—such as high-performance liquid chromatography (HPLC), high-performance thin-layer chromatography (HPTLC), and supercritical-fluid chromatography (SFC)—are improved variations of either adsorption or partitioning methods.

Column Chromatography Open-bed columns of various materials have been used for over 40 years. Silica gel has been used for the fractionation of air particles (Wynder and Hoffmann, 1965) and cigarette-smoke condensates (Kier et al., 1974; Lee et al., 1977). Airborne particles have also been fractionated on alumina (Dehnen et al., 1977; Tokiwa et al., 1977). Solvent partitioning, Sephadex chromatography, and alumina chromatography (in the case of separations of fractions from synfuels) have been reported (Guerin et al., 1980; Hsie et al., 1980).

Chromatography with alumina, Florisil, XAD-2, Ambersorb, and silica gel has been evaluated for separating organic substances from drinking water

(Chriswell et al., 1978). Because of poor recovery and the production of chemical-reaction artifacts of organic compounds during fractionation with silica or alumina, Florisil might be preferred in some circumstances. Artifacts can often be minimized and better reproducibility attained when the adsorptive activity of the sorbent is controlled. It is important to define the state of activation of the medium.

Gel-Permeation Chromatography Because of the potential for artifacts of susceptible compounds in acid-base fractionation schemes, alternate schemes with gel-permeation chromatography (GPC) have been developed (Guerin et al., 1978; Epler et al., 1978). This technique has been very beneficial for separating PAHs. Molecular size, shape, and polarity are major influences on the separation of PAHs (Wilk et al., 1966; Edstrom and Petro, 1968; Oelert, 1969; Vithayathil et al., 1978).

GPC methods are believed to involve hydrophobic-hydrophilic partitioning, molecular size separation, and aliphatic-aromatic separation. Basic fractions can also be separated with Sephadex chromatography (Guerin et al., 1978). Serum extracts of fly ash have been subjected to GPC to separate protein-bound mutagens from low-molecular-weight mutagens (Chrisp et al., 1978).

Thin-Layer Chromatography Thin-layer chromatography (TLC) is generally considered a mild method and can be used for especially sensitive compounds. TLC has been applied not only to air particles, but to coal-derived heavy distillates (Wilson et al., 1980), diesel particulate extracts (Pederson and Siak, 1981; Pederson, 1983), and other complex mixtures.

High-Performance Liquid Chromatography Because HPLC has the advantages of higher reproducibility and shorter analysis time, it has for the most part supplanted open-column chromatography and TLC. However, HPLC and TLC are complementary techniques; HPLC can be used to measure accurately the components identified by TLC with its impressive array of specific colorimetric sprays. Both normal-phase chromatography and reverse-phase chromatography are widely used (Eisenberg, 1978; Kaiser et al., 1981; Rappaport et al., 1980; Later and Lee, 1983; Later et al., 1981; Toste et al., 1982). Because the mobile phase can interact with the solute at mild operating temperatures, HPLC is suitable for the analysis of nonvolatile and thermally labile compounds (Ahuja, 1984). Nonetheless, achieving good resolution of these types of polar analytes with HPLC is still a problem. This is a concern, because the polar fraction is often also toxic.

High-Resolution Techniques High- and ultrahigh-resolution techniques, currently under vigorous development, are the analytic tools of the future for resolving complex mixtures, particularly those with many components in the same chemical class. High-resolution techniques are not practical for generating fractions for bioassay (the quantities obtained are insufficient for most assays), so a search for causative agents generally stops with subfractions that might still be relatively complex. If toxicity is associated with subfractions that are prepared by the above methods, then high-resolution chromatography is applied in concert with chemical analysis. If all or most of the analytes in a subfraction are identified, then a predictive strategy can be applied to the individual components found to be present.

High-resolution chromatography evolved with capillary gas chromatography—a technique that remained dormant for a long period, because it was difficult to prepare suitable glass capillary columns, but changed abruptly with the introduction of fused-silica capillaries (Ahuja, 1984). High-resolution gas-liquid chromatography is often used in combination with a detector that yields molecular information (e.g., mass spectrometry and Fourier-transform infrared spectroscopy).

Even though excellent separating power is attainable with capillary gas-liquid chromatography, only 10% of the millions of known compounds are amenable to this technique (Gouw et al., 1979). The unequivocal separation of all constituents in very complex chemical mixtures, such as synfuels, is rarely achieved (Later and Wright, 1984). The limitations stem from two factors—analyte volatility and polarity. Chemicals with low volatility (e.g., boiling point over 350°C) or high polarity will not go through a gas-liquid chromatographic column. That is a serious limitation, because toxic chemicals do not have the same polarity and volatility boundaries. Clearly, the separation of analytes in complex mixtures requires exquisite resolution, if the specific chemicals responsible for toxicity in mixtures are to be assessed comprehensively. Two-dimensional systems, in which a portion of the eluent is rechromatographic under different conditions, have also shown promise (C. W. Wright, 1984).

Research progress has been steady in high-resolution separation techniques, and complementary methods, including computerized data analysis, have now been developed that allow the remaining chemicals to be analyzed (Demirgian, 1984; Stamoudis and Demirgian, 1985; Stamoudis, 1982).

HPLC has made substantial advances in the last decade (Ishii and Takeuchi, 1984). New avenues for achieving ultrahigh resolution have been developed, and micropacked fused silica for microbore columns and long capillary columns can provide an enormous number of theoretical plates (that is, a number of degrees of separation equivalent to many simple distillations) and high resolution that was not possible a few years ago (Ishii and Takeuchi, 1984). Analytes with widely differing polarity can be separated on high-resolution micropacked fused-silica columns in combination with gradient elution. Microdetectors have also been designed—a requisite in combination with high resolution. HPLC is more versatile than GLC, but has been difficult to use with molecular detection systems.

Supercritical-fluid chromatography (SFC) takes advantage of the special properties of substances that have been compressed beyond and heated above their critical pressures and temperatures (Eckert et al., 1986). This emerging technique provides yet another dimension to ultrahigh-resolution chromatography and permits resolution of polar chemicals that is not possible with GLC or, in some cases, even with HPLC (Randall, 1984; Novotny, 1985). Because of high solute diffusivity, lower viscosity, and excellent solvating properties, high resolution can be attained. Because the supercritical fluid used as the mobile phase is a hybrid with some characteristics of both gases and liquids (*n*-pentane, carbon dioxide, and nitrous oxide), it can be considered as a combination of and complementary to gas chromatography (GC) and HPLC (Randall, 1984). High chromatographic efficiencies and fast analysis, compared with those of HPLC, are predominant characteristics. Nonvolatile and high-molecular-weight chemicals can be separated at relatively low temperatures with efficiencies approaching those of GLC. Relatively polar analytes can be separated effectively (Randall, 1984). Current results indicate its great potential for separation of complex mixtures (Randall, 1984; Chess and Smith, 1984; Smith et al., 1984; Wright et al., 1984). The high resolution obtainable and the ability to couple capillary SFC to most of the conventional HPLC and GC detectors make this technique very versatile. Detectors that have been successfully coupled to SFC include ultraviolet-absorption, fluorescence, flame-ionization, and nitrogen thermoionic detectors, as well as mass spectrometry. The available methods of separating complex mixtures are HPTLC, SFC, field-flow fractionation, electrophoresis, electroosmosis, and isoelectric focusing. When choosing a chromatographic technique for separating complex mixtures, the investigator must consider analyte volatility and thermal stability, chromatographic efficiency, and speed of analysis. Development of a strategy, however, has not been explicitly standardized.

Chemical Analyses

An extensive repertoire of analytic techniques (Table 4-4) are available for structural elucidation of organic molecules. The terms "spectroscopy" and "spectrometry" encompass a group of techniques that differ widely in their mode of application and in the information they reveal about chemical structure (Buchanan et al., 1983). A comprehensive strategy has not been devised for applying spectroscopic techniques to the analysis of organic substances and complex mixtures. The approach selected generally is targeted to obtain information about a specific portion of a mixture, but a comprehensive strategy would be useful in extracting the relevant molecular information that can be

TABLE 4-4 Conventional Instrumental Techniques for Chemical Analysis

Instrumental Technique	Information Content	Selected References
Ultraviolet and visible spectroscopy	Conjugation	Brown, 1980; Silverstein and Bassler, 1963
Vibrational spectroscopy (infrared and Raman)	Chemical functionalities and assembly of atoms	Borwn, 1980; Strommen and Nakamoto, 1984; Gans, 1980; Case and Fately, 1980; Clerc et al., 1981
Molecular emission spectroscopy	Selective detection	Penzer, 1980
Mass spectrometry	Molecular weight, chemical functionalities, overall structure, and assembly of atoms	Watson and Throck, 1985; Wilson et al., 1981; Smith and Udseth, 1983
Nuclear magnetic resonance spectrometry	Arrangement of atoms in molecule	Jones, 1980; Jackman and Sternhell, 1969
Atomic absorption spectroscopy and inductively coupled argon plasma emission	Elements	Wood, 1980; Cantle, 1982
Electron paramagnetic spin resonance	Free-radical formation	Brown, 1980

used to identify the chemicals responsible for toxicity. Extreme caution should be exercised when one is comparing chemical data bases derived from different complex mixtures, because the separation techniques introduce uncertainty as to whether structural elucidation is unequivocal or merely postulative or tentative.

For structural analysis, these modern methods are used most efficiently if they are combined, because they can provide complementary information that increases their overall effectiveness. Guidelines for the combined application of spectroscopic methods are available (Clerc et al., 1981; Silverstein and Bassler, 1963). The most generally applied methods for the characterization of organic substances are ultraviolet spectroscopy, infrared spectroscopy, proton and carbon-13 nuclear magnetic resonance, and mass spectrometry (Clerc et al., 1981). Specialized techniques—such as magnetic resonance of other nuclei, Raman spectroscopy, and optical rotary dispersion—are used less often. Gas chromatography/Fourier-transform infrared (GC/FT-IR) spectroscopy is now being developed for routine use in identifying analytes in complex mixtures with the aid of software programs for the automatic interpretation of spectra (Griffiths, 1980; Gurka, 1985). Infrared, Raman, visible, ultraviolet, electron-spin resonance, and magnetic circular dichroism spectroscopies have all been used in conjunction with matrix isolation (Barnes and Orville-Thomas, 1980). Matrix isolation traps isolated molecules of the species to be studied in a large excess of an inert material by rapid condensation at a low temperature, so that the diluent forms a rigid cage or matrix (Barnes and Orville-Thomas, 1980). The noble gases (primarily argon) and nitrogen are most widely used for matrix materials. More recent uses of matrix isolation have been in combination with GC effluents in which high-resolution gas chromatography (HRGC) has been coupled with FT-IR spectroscopy. Several HPLC/FT-IR systems have been described (Griffiths, 1980). The exquisite sensitivity of GC/FT-IR with matrix isolation rivals that of mass spectrometry; these two methods can also be combined to provide simultaneous and complementary information during chromatography of a complex mixture.

The gas chromatograph is the most commonly used inlet system for analysis of complex mixtures with mass spectrometry (Watson and Throck, 1985). It is rarely possible to isolate in pure form each of the hundreds of individual substances present. However, HRGC combined with mass spectrometry (GC/MS) can often serve as the final purification step by resolving the various components and presenting them one at a time to the mass spectrometer.

The mass spectrometer can also serve as a universal detector for SFC and HPLC (Smith and Udseth, 1983). Several types of devices are under development (e.g., moving wire or moving belt, direct inlet, and thermospray) to introduce relatively nonvolatile materials after separation into the ion source of the mass spectrometer. A fast-atom-bombardment (FAB) ionization technique combined with a moving belt exploits the attributes of FAB ionization of nonvolatile analytes and allows analysis in a manner very similar to that of GC/MS (Smith and Udseth, 1983).

An alternative to resolving components chromatographically is separating them by mass spectrometric methods—the so-called MS/MS methods (Watson and Throck, 1985; Smith and Udseth, 1983; Henderson et al., 1982, 1983, 1984). The complex mixture is ionized, usually gently, to produce characteristic ions from each component. The ions suspected to arise from the component of interest are selected by mass analysis in the first stage of a double mass spectrometer, and their identity is confirmed by mass analysis in the second stage on the basis of the fragment ions arising from collision-induced association of the selected ions.

REFERENCES

Ahuja, S. 1984. Overview: Multiple pathways to ultrahigh resolution chromatography, pp. 1–8. In S. Ahuja (ed.). Ultrahigh Resolution Chromatography. ACS Symposium Series 250. American Chemical Society, Washington, D.C.

Alfheim, I., G. Löfroth, and M. Moller. 1983. Bioassay of extracts of ambient particulate matter. Environ. Health Perspect. 47:227–238.

Alfheim, I., A. Bjorseth, and M. Moller. 1984. Characterization of microbial mutagens in complex samples—methodology and application. CRC Crit. Rev. Environ. Control 14:91–150.

Baker, R. A. 1970. Trace organic contaminant concentration by freezing. IV: Ionic effects. Water Res. 4:559–573.

Barnes, A. J., and W. J. Orville-Thomas. 1980. FT-IR matrix isolation studies, pp. 157–170. In J. R. Durig (ed.). Analytical Applications of FT-IR to Molecular and Biological Systems. D. Reidal, Dordrecht, Holland. (607 pp.)

Barnhart, B. J., and S. H. Cox. 1980. Mutation of Chinese hamster cells by near-UV activation of promutagens. Mutat. Res. 72:135–142.

Benson, J. M., R. L. Hanson, R. E. Royer, C. R. Clark, and R. F. Henderson. 1984. Toxicological and chemical characterization of the process stream materials and gas combustion products of an experimental low-Btu coal gasifier. Environ. Res. 33:396–412.

Berkley, R. E., and E. D. Pellizzari. 1978. Evaluation of Tenax GC sorbent for *in situ* formation of N-nitrosodimethylamine. Anal. Lett. 4:327–346.

Bock, F. G., A. P. Swain, and R. L. Stedman. 1969. Bioassay of major fractions of cigarette smoke condensate by an accelerated technic. Cancer Res. 29:584–587.

Boparai, A. S., D. A. Haugen, K. M, Suhrbier, and J. F. Schneider. 1983. An improved procedure for extraction of aromatic bases from synfuel materials, pp. 3–11. In C. W. Wright, W. C. Weimer, and W. D. Felix (eds.). Advanced Techniques in Synthetic Fuels Analysis. PNL-SA-11552. CONF-811160. U.S. Department of Energy, Technical Information Center, Oak Ridge, Tenn. (Available from NTIS as DE83015528.)

Bridbord, K., and J. G. French. 1978. Carcinogenic and mutagenic risks associated with fossil fuels, pp. 451–463. In P. W. Jones and R. I. Freudenthal (eds.). Carcinogenesis—A Comprehensive Survey, Vol. 3. Polynuclear Aromatic Hydrocarbons. Raven Press, New York.

Brown, K. W., K. C. Donnelly, J. C. Thomas, P. Davol, and B. R. Scott. 1985. Mutagenicity of three agricultural soils. Sci. Total Environ. 35:799–807.

Brown, S. B. 1980. Introduction to spectroscopy, pp. 1–13. In S. B. Brown (ed.). An Introduction to Spectroscopy for Biochemists. Academic Press, New York.

Browning, D. R. 1969. Chromatography. McGraw-Hill, New York. (151 pp.)

Brusick, D. J., and R. R. Young. 1982. IERL-RTP Procedures Manual: Level 1 Environmental Assessment, Biological Tests. EPA 600/S8-81-024. Office of Research and Development, U.S. Environmental Protection Agency, Washington, D.C. (8 pp.)

Buchanan, M. V., M. R. Guerin, G. L. Kao, I. B. Rubin, and J. E. Caton. 1983. Comparative spectroscopic characterization of synthetic fuels, pp. 286–298. In C. W. Wright, W. C. Weimer, and W. D. Felix (eds.). Advanced Techniques in Synthetic Fuels Analysis. PNL-SA-11552. CONF-811160. U.S. Department of Energy, Technical Information Center, Oak Ridge, Tenn. (Available from NTIS as DE83015528.)

Burke, F. P., R. A. Winschel, and G. A. Robbins. 1984. Recycle Slurry Oil Characterization: Third Annual Report. DOE/PC 30027-56. Conoco Inc., Coal Research Division, Library, Pa.

Burton, F. G., R. E. Schirmer, D. D. Mahlum, and F. D. Andrew. 1979. Disposition of process solvent from solvent refined coal in tissues of the rat after oral dosing. Abstract No. 54. Toxicol. Appl. Pharmacol. 48:A27.

Cantle, J. E. 1982. Atomic Absorption Spectrometry. Elsevier, New York. (448 pp.)

Carey, A. E., and J. A. Gowen. 1978. PCB's in Agricultural and Urban Soil. U.S. Environmental Protection Agency, Washington, D.C. (Available from NTIS as PB-276 315/9.) (4 pp.)

Case, J. C., and W. G. Fately. 1980. One view of the advantages of infrared interferometry, pp. 3–10. In J. R. Durig (ed.). Analytical Applications of FT-IR to Molecular and Biological System. D. Reidel, Dordrecht, Holland. (607 pp.)

Cerniglia, C. E. 1981. Aromatic hydrocarbons: Metabolism by bacteria, fungi, and algae, pp. 321–361. In E. Hodgson, J. R. Bend, and R. M. Philpot (eds.). Reviews in Biochemical Toxicology. Vol. 3. Elsevier, New York.

Chan, T. L., P. S. Lee, and J.-S. Siak. 1981. Diesel-particulate collection for biological testing: Comparison of electrostatic precipitation and filtration. Environ. Sci. Technol. 15:89–93.

Cheng, Y.-S., and H.-C. Yeh. 1979. Particle bounce in cascade impactors. Environ. Sci. Technol. 13:1392–1396.

Cheng, Y.-S., R. L. Hanson, R. L. Carpenter, C. H. Hobbs. 1984. Use of a massive volume air sampler to collect fly ash for biological characterization. J. Air Pollut. Control Assoc. 34:671–674.

Chess, E. K., and R. D. Smith. 1984. Development and Evaluation of Supercritical Fluid Chromatography/Mass Spectrometry for Polar and High Molecular Weight Coal Components. PNL-SA-12298. CONF-840694-9. Battelle Pacific Northwest Labs, Richland, Wash. (Available from NTIS as DE85000827/XAB.) (8 pp.)

Chrisp, C. E., and G. L. Fisher. 1980. Mutagenicity of airborne particles. Mutat. Res. 76:143–164.

Chrisp, C. E., G. L. Fisher, and J. E. Lammert. 1978. Mutagenicity of filtrates from respirable coal fly ash. Science 199:73–75.

Chriswell, C. D., B. A. Glatz, J. S. Fritz, and H. J. Svec. 1978. Mutagenic analysis of drinking water, pp. 477–494. In M. D. Waters, S. Nesnow, J. F. Huisingh, S. S. Sandhu, and L. Claxton (eds.). Application of Short-Term Bioassays in the Fractionation and Analysis of Complex Environmental Mixtures. Plenum, New York.

Cifrulak, S. D. 1969. Spectroscopic evidence of phthalates in soil organic matter. Soil Sci. 107:63–69.

Clark, C. R., T. J. Truex, F. S. C. Lee, and I. T. Salmeen. 1981. Influence of sampling filter type on the mutagenicity of diesel exhaust particulate extracts. Atmos. Environ. 15:397–402.

Claxton, L. D. 1982. Review of fractionation and bioassay characterization techniques for the evaluation of organics associated with ambient air particles, pp. 19–34. In R. R. Tice, D. L. Costa, and K. M. Schaich (eds.). Genotoxic Effects of Airborne Agents. Environmental Science Research. Vol. 25. Plenum, New York.

Clerc, J. T., E. Pretsch, and J. Seibl. 1981. Structural Analysis of Organic Compounds by Combined Application of Spectroscopic Methods. Elsevier, New York. (288 pp.)

Cochran, W. G. 1977. Sampling Techniques, 3rd ed. John Wiley & Sons, New York. (274 pp.)

Corn, M., T. L. Montgomery, and N. A. Esmen. 1971. Suspended particulate matter: Seasonal variation in specific surface areas and densities. Environ. Sci. Technol. 5:155–158.

Crathorne, B., C. D. Watts, and M. Fielding. 1979. Analysis of non-volatile organic compounds in water by high-performance liquid chromatography. J. Chromatography 185:671–690.

Daisey, J. M., T. J. Kneip, I. Hawryluk, and F. Mukai. 1980. Seasonal variations in the bacterial mutagenicity of airborne particulate organic matter in New York City. Environ. Sci. Technol. 14:1487–1490.

Dawson, R., and K. Mopper. 1978. A note on the losses of monosaccharides, amino sugars, and amino acids from extracts during concentration procedures. Anal. Biochem. 84:186–190.

Dehnen, W., N. Pitz, and R. Tomingas. 1977. The mutagenicity of airborne particulate pollutants. Cancer Lett. 4:5–12.

Demirgian, J. 1984. Computerized rapid analysis of complex mixtures by gas chromatography. J. Chromatogr. Sci. 22:153–160.

De Wiest, F., and D. Rondia. 1976. Sur la validité des déterminations du Benzo (a) pyrène atmosphérique pendant les mois d'été. Atmos. Environ. 10:487–489. (English abstract.)

Eckert, C. A., J. G. Van Alsten, and T. Stoicos. 1986. Supercritical fluid processing. Environ. Sci. Technol. 20:319–325.

Edstrom, T., and B. A. Petro. 1968. Gel permeation chromatographic studies of polynuclear aromatic hydrocarbon materials. J. Polym. Sci. Part C (21):171–182.

Eisenberg, W. C. 1978. Fractionation of organic material extracted from suspended air particulate matter using high pressure liquid chromatography. J. Chromatogr. Sci. 16:145–151.

Epler, J. L. 1980. The use of short-term tests in the isolation and identification of chemical mutagens in complex mixtures, pp. 239–270. In F. J. de Serres and A. Hollaender (eds.). Chemical Mutagens: Principles and Methods for Their Detection, Vol. 6. Plenum, New York.

Epler, J. L., B. R. Clark, C.-H. Ho, M. R. Guerin, and T. K. Rao. 1978. Short-term bioassay of complex organic mixtures: Part II, mutagenicity testing, pp. 269–289. In M. D. Waters, S. Nesnow,

J. L. Huisingh, S. S. Sandhu, and L. Claxton (eds.). Application of Short-Term Bioassays in the Fractionation and Analysis of Complex Environmental Mixtures. Plenum, New York.

Fitz, D. R., G. J. Doyle, and J. N. Pitts, Jr. 1983. An ultrahigh volume sampler for the multiple filter collection of respirable particulate matter. J. Air Pollut. Control Assoc. 33:877–879.

Flessel, P., G. Guirguis, J. Cheng, K. Chang, and E. Hahn. 1984. Monitoring of Mutagens and Carcinogens in Community Air. ARB-R-84/223. California State Air Resources Board, Sacramento, Calif. (Available from NTIS as PB85-173763/XAB.) (134 pp.)

Florin, I., L. Rutberg, M. Curvall, and C. R. Enzell. 1980. Screening of tobacco smoke constituents for mutagenicity using the Ames' test. Toxicology 15:219–232.

Flotard, R. D. 1980. Sampling and Analysis of Trace-Organic Constituents in Ambient and Workplace Air at Coal-Conversion Facilities. Argonne National Laboratory, Argonne, Ill. (Available from NTIS as ANL/PAG-3.) (51 pp.)

Gans, P. 1980. Vibrational spectroscopy, pp. 115–147. In S. B. Brown (ed.). An Introduction to Spectroscopy for Biochemists. Academic Press, New York.

Gilbert, R. O. 1983. Field sampling designs, simple random and stratified random sampling. TRANS-STAT Statistics for Environmental Studies, No. 24. PNL-SA-11551. Battelle Pacific Northwest Labs, Richland, Wash. (Available from NTIS as DE83016826.) (38 pp.)

Gilbert, R. O. 1984. Field sampling designs: Systematic sampling. TRANS-STAT Statistics for Environmental Studies. No. 26. PNL-SA-12180. Battelle Pacific Northwest Labs, Richland, Wash. (32 pp.)

Göggelmann, W., and P. Spitzauer. 1982. Mutagenic activity, content of polycyclic aromatic hydrocargons (PAH) and humus in agricultural soils. Abstract No. 50. Mutat. Res. 97:189–190.

Gorse, R. A., Jr., I. T. Salmeen, and C. R. Clark. 1982. Effects of filter loading and filter type on the mutagenicity and composition of diesel exhaust particulate extracts. Atmos. Environ. 16:1523–1528.

Gouw, T. H., R. E. Jentoft, and E. J. Gallegos. 1979. Some recent advances in supercritical fluid chromatography, pp. 583–592. In K. D. Timmerhaus and M. S. Barber (eds.). High-Pressure Science and Technology, Vol. 1: Physical Properties and Material Synthesis. Plenum, New York. (583 pp.)

Griest, W. H., C. E. Higgins, R. W. Holmberg, J. H. Moneyhun, J. E. Caton, J. S. Wike, and R. R. Reagen. 1982. Characterization of vapor- and particulate-phase organics from ambient air sampling at the Kosovo gasifier, pp. 395–410. In L. H. Keith (ed.). Energy and Environmental Chemistry, Vol. 1. Ann Arbor Science, Ann Arbor, Mich.

Griffiths, P. R. 1980. Chromatography and FT-IR spectrometry, pp. 149–155. In J. R. Durig (ed.). Analytical Applications of FT-IR to Molecular and Biological Systems. D. Reidal, Dordrecht, Holland. (607 pp.)

Guerin, M. R. 1981. The integrated approach to chemical-biological analysis, pp. 1–16. In J. C. Harris, P. L. Levins, and K. D. Drewitz (eds.). Proceedings, 2nd Symposium on Process Measurements for Environmental Assessment, February 25–27, Atlanta, Ga., 1980. Industrial Environmental Research Lab., Research Triangle Park, N.C. (Available from NTIS as PB82-211574.)

Guerin, M. R., B. R. Clark, C.-H. Ho, J. L. Epler, and T. K. Rao. 1978. Short-term bioassay of complex organic mixtures: Part I, chemistry, pp. 247–268. In M. D. Waters, S. Nesnow, J. L. Huisingh, S. S. Sandhu, and L. Claxton (eds.). Application of Short-Term Bioassays in the Fractionation and Analysis of Complex Environmental Mixtures. Plenum, New York.

Guerin, M. R., C.-H. Ho, T. K. Rao, B. R. Clark, and J. L. Epler. 1980. Separation and identification of mutagenic constituents of petroleum substitutes. Int. J. Environ. Anal. Chem. 8:217–225.

Guerin, M. R., J. Dutcher, E. S. Olson, E. J. Peterson, V. C. Stamoudis, D. H. Stuermer, and B. W. Wilson. 1983. Summary of chemical characterization methodologies with future directives, pp. 263–266. In C. W. Wright, W. C. Weimer, and W. D. Felix (eds.). Advanced Techniques in Synthetic Fuels Analysis. PNL-SA-11552. CONF-811160. U.S. Department of Energy, Technical Information Center, Oak Ridge, Tenn. (Available from NTIS as DE83015528.)

Gurka, D. F. 1985. Interim protocol for the automated analysis of semivolatile organic compounds by gas chromatography/Fourier transform infrared (GC/FT-IR) spectrometry. Appl. Spectrosc. 39:827–833.

Hackett, P. L., R. L. Music, D. D. Mahlum, and M. R. Sikov. 1983. Developmental effects of oral administration of solvent refined coal materials to rats. (Abstract.) Teratology 27:47A.

Hanson, R. L., R. L. Carpenter, and G. J. Newton. 1980. Chemical characterization of polynuclear aromatic hydrocarbons in airborne effluents from an experimental fluidized bed combustor, pp. 599–616. In A. Bjørseth and A. J. Dennis (eds.). Polynuclear Aromatic Hydrocarbons: Chemistry and Biological Effects. Battelle Press, Columbus, Ohio.

Hanson, R. L., C. R. Clark, R. L. Carpenter, and C. H. Hobbs. 1981. Evaluation of Tenax-GC and XAD-2 as polymer adsorbents for sampling fossil fuel combustion products containing nitrogen oxides. Environ. Sci. Technol. 15:701–705.

Hanson, R. L., R. E. Royer, J. M. Benson, R. L. Carpenter, G. J. Newton and R. F. Henderson. 1982. Chemical fractionation and analysis of organic compounds in process streams of low Btu gasifier effluents, pp. 205–223. In E. L. Fuller, Jr. (ed.). Coal and Coal Products: Analytical Characterization Techniques. Polynuclear Aromatic Hydrocarbons. ACS Symposium Series 205. American Chemical Society, Washington, D.C.

Hanson, R. L., C. R. Clark, R. L. Carpenter, and C. H. Hobbs. 1984. Comparison of Tenax-GC and XAD-2 as polymer adsorbents for sampling combustion exhaust gases, pp. 79–93. In L. H. Keith (ed.). Identification and Analysis of Organic Pollutants in Air. Butterworth, Boston, Mass.

Hanson, R. L., A. R. Dahl, S. J. Rothenberg, J. M. Benson, A. L. Brooks, and J. S. Dutcher. 1985. Chemical and biological characterization of volatile components of environmental samples after fractionation by vacuum line cryogenic distillation. Arch. Environ. Contam. Toxicol. 14:289–297.

Haugen, D. A., and V. C. Stamoudis. 1986. Isolation and identification of mutagenic polycyclic aromatic hydrocarbons from a coal gasifier condensate. Environ. Res. 41:400–419.

Henderson, T. R., R. E. Royer, C. R. Clark, T. M. Harvey, and D. F. Hunt. 1982. MS/MS of diesel emissions and fuels treated with NO_2. J. Appl. Toxicol. 2:231–237.

Henderson, T. R., J. D. Sun, R. E. Royer, C. R. Clark, A. P. Li, T. M. Harvey, D. H. Hunt, J. E. Fulford, A. M. Lovette, and W. R. Davidson. 1983. Triple quadrupole mass spectrometry studies of nitroaromatic emissions from different diesel engines. Environ. Sci. Technol. 17:443–449.

Henderson, T. R., J. D. Sun, A. P. Li, R. L. Hanson, W. E. Bechtold, T. M. Harvey, J. Shabanowitz, and D. F. Hunt. 1984. GC/MS and MS/MS studies of diesel exhaust mutagenecity and emissions from chemically defined fuels. Environ. Sci. Technol. 18:428–434.

Henry, W. M., R. I. Mitchell, and R. J. Thompson. 1978. Development of a Large Sampler Collector of Respirable Particulate Matter. EPA-600/4-78/009. Battelle Columbus Labs, Columbus, Ohio. (Available from NTIS as PB-281 528/0.) (54 pp.)

Higgins, C. E., W. H. Griest, and G. Olerich. 1983. Application of Tenax trapping to analysis of gas phase organic compounds in ultra-low tar cigarette smoke. J. Assoc. Off. Anal. Chem. 66:1074–1083.

Higgins, C. E., W. H. Griest, and M. R. Guerin. 1984. Sampling and Analysis of Cigarette Smoke Using the Solid Adsorbent Tenax. ORNL/TM-9167. Oak Ridge National Lab, Oak Ridge, Tenn. (Available from NTIS as DE 84012025.) (25 pp.)

Hite, M., and H. Skeggs. 1979. Mutagenic evaluation of nitroparaffins in Salmonella typhimurium/mammalian-microsome test and the micronucleus test. Environ. Mutagen. 1:383–389.

Ho, C.-H., C. Y. Ma, B. R. Clark, M. R. Guerin, T. K. Rao, and J. L. Epler. 1980. Separation of neutral nitrogen compounds from synthetic crude oils for biological testing. Environ. Res. 22: 412–422.

Hoffmann, D., K. Norpoth, R. H. Wickramasinghe, and G. Müller. 1980. The detection of mutagenic air pollutants from filter samples by the salmonella/mammalian S-9 mutagenicity test (Ames test) with S. typhimurium TA98 (Part 1). Zentralbl. Bakteriol. Mikrobiol. Hyg. 1 Abt., Orig. B 171: 388–407.

Hsie, A. W., P. A. Brimer, J. P. O'Neill, J. L. Epler, M. R. Guerin, and M. H. Hsie. 1980. Mutagenicity of alkaline constituents of a coal-liquefied crude oil in mammalian cells. Mutat. Res. 78:79–84.

Hughes, T. J., E. Pellizzari, L. Little, C. Sparacino, and A. Kolber. 1980. Ambient air pollutants: Collection, chemical characterization and mutagenicity testing. Mutat. Res. 76:51–83.

ISO (International Organization for Standardization). 1983. Air Quality—Particle Size Fraction Definitions for Health-Related Sampling. ISO/TR 7708-1983. International Organization for Standardization, Geneva. (13 pp.)

Ishii, D., and T. Takeuchi. 1984. Application of micro high performance liquid chromatography to the separation of complex mixtures, pp. 109–120. In S. Ahuja (ed.). Ultrahigh Resolution Chromatography. ACS Symposium Series 250. American Chemical Society, Washington, D.C.

Jackman, L. M., and S. Sternhell. 1969. Applications of Nuclear Magnetic Resonance Spectroscopy in Organic Chemistry. 2nd ed. Pergamon, New York. (456 pp.)

Jensen, V. 1975. Bacterial flora of soil after application of oily waste. Oikos 26:152–158.

Jobson, A., M. McLaughlin, F. D. Cook, and D. W. S. Westlake. 1974. Effect of amendments on the microbial utilization of oil applied to soil. Appl. Microbiol. 27:166–171.

Jolley, R. L. 1981. Concentrating organics in water for biological testing. Environ. Sci. Technol. 15:874–880.

Jolley, R. L., S. Katz, and J. E. Mrochek. 1975. Analyzing organics in complex, dilute aqueous solutions. Chem. Technol. 5:312–318.

Jones, R. 1980. Nuclear magnetic resonance, pp. 235–278. In S. B. Brown (ed.). An Introduction to Spectroscopy for Biochemists. Academic Press, New York.

Kaiser, C., A. Kerr, D. R. McCalla, J. N. Lockington, and E. S. Gibson. 1981. Use of bacterial mutagenicity assays to probe steel foundry lung cancer hazard, pp. 583–592. In M. Cooke and A. J. Dennis (eds.). Chemical Analysis and Biological Fate: Polynuclear Aromatic Hydrocarbons. Battelle Press, Columbus, Ohio.

Kalkwarf, D. R., D. L. Steward, R. A. Pelroy, and W. C. Weimer. 1984. Photodegradation of Mutagens in Solvent-Refined Coal Liquids. PNL-4982. Battelle Pacific Northwest Labs, Richland, Wash. (Available from NTIS as DE84010760.) (46 pp.)

Kelman, B. J., and D. L. Springer. 1981. Fetal exposure to benzo[a]pyrene across the hemochorial placenta, pp. 387–397. In D. D. Mahlum, R. H. Gray, and W. D. Felix (eds.). Coal Conversion and the Environment: Chemical, Biomedical, and Ecological Considerations. Proceedings of the Twentieth Hanford Life Sciences Symposium at Richland, Wash., October 19–23, 1980. CONF-801039. U.S. Department of Energy, Technical Information Center, Oak Ridge, Tenn. (Available from National Technical Information Service as DE82000105.)

Kier, L. D., E. Yamasaki, and B. N. Ames. 1974. Detection of mutagenic activity in cigarette smoke condensates. Proc. Nat. Acad. Sci. U.S.A. 71:4159–4163.

Kinae, N., T. Hashizume, T. Makita, I. Tomita, and I. Kimura. 1981. Kraft pulp mill effluent and sediment can retard development and lyse sea urchin eggs. Bull. Environ. Contam. Toxicol. 27: 616–623.

Kool, H. J., C. F. van Keijl, H. J. van Kranen, and E. de Greef. 1981a. The use of XAD-resins for the detection of mutagenic activity in water. I. Studies with surface water. Chemosphere 10:85–98.

Kool, H. J., C. F. van Keijl, H. J. van Kranen, and E. de Greef. 1981b. The use of XAD-resins for the detection of mutagenic activity in water. II. Studies with drinking water. Chemosphere 10:99–108.

Kopfler, F. C. 1980. Alternative strategies and methods for concentrating chemicals from water, pp. 141–153. In M. D. Waters, S. S. Sandhu, J. L. Huisingh, L. Claxton, and S. Nesnow (eds.). Short-Term Bioassays in the Analysis of Complex Environmental Mixtures. II. Plenum, New York.

Kopfler, F. C., W. E. Coleman, R. G. Melton, R. G. Tardiff, S. C. Lynch, and J. K. Smith. 1977. Extraction and identification of organic micropollutants: Reverse osmosis method. Ann. N.Y. Acad. Sci. 298:20–30.

Kringstad, K. P., P. O. Ljungquist, F. de Sousa, and L. M. Strömberg. 1981. Identification and mutagenic properties of some chlorinated aliphatic compounds in the spent liquor from kraft pulp chlorination. Environ. Sci. Technol. 15:562–566.

Krost, K. J., E. D. Pellizzari, S. G. Walburn, and S. A. Hubbard. 1982. Collection and analysis of hazardous organic emissions. Anal. Chem. 54:810–817.

Larson, R. A., L. L. Hunt, and D. W. Blankenship. 1977. Formation of toxic products from a #2 fuel oil by photooxidation. Environ. Sci. Technol. 11:492–496.

Later, D. W., and M. L. Lee. 1983. Chromatographic methods for the chemical and biological characterization of polycyclic aromatic compounds in synfuel materials, pp. 44–73. In C. W. Wright, W. C. Weimer, and W. D. Felix (eds.). Advanced Techniques in Synthetic Fuels Analysis. PNL-SA-11552. CONF-811160. U.S. Department of Energy, Technical Information Center, Oak Ridge, Tenn. (Available from NTIS as DE83015528.)

Later, D. W., and B. W. Wright. 1984. Capillary column gas chromatographic separation of amino polycyclic aromatic hydrocarbon isomers. J. Chromatography 289:183–193.

Later, D. W., M. L. Lee, K. D. Bartle, R. C. Kong, and D. L. Vassilaros. 1981. Chemical class separation and characterization of organic compounds in synthetic fuels. Anal. Chem. 53:1612–1620.

Later, D. W., T. G. Andros, and M. L. Lee. 1983a. Isolation and identification of amino polycyclic aromatic hydrocarbons from coal-derived products. Anal. Chem. 55:2126–2132.

Later, D. W., C. W. Wright, and B. W. Wilson. 1983b. The analytical chemistry of products from process strategies designed to reduce the biological activity of direct coal liquefaction materials. Am. Chem. Soc., Div. Fuel Chem. Preprints 28(5):273–284.

Lee, P. N., K. Rothwell, and J. K. Whitehead. 1977. Fractionation of mouse skin carcinogens in cigarette smoke condensate. Br. J. Cancer 35:730–742.

Lentzen, D. W., D. E. Wagoner, E. D. Estes, and W. F. Gutknecht. 1978. IERL-RTP Procedures Manual: Level 1 Environmental Assessment. 2nd ed. EPA/600/7-78/201. Industrial Environmental Research Lab, Research Triangle Park, N.C. (Available from NTIS as PB-293 795/1.) (279 pp.)

Lippmann, M. 1985. Development of particle size-selective threshold limit values. Ann. Am. Conf. Ind. Hyg. 12:27–34.

Mahlum, D. D. 1983a. Initiation/promotion studies with coal-derived liquids. J. Appl. Toxicol. 3:31–34.

Mahlum, D. D. 1983b. Skin-tumor initiation activity of coal liquids with different boiling-point ranges. J. Appl. Toxicol. 3:254–258.

Mahlum, D. D., and D. L. Springer. 1986. Teratogenic response of the rat and mouse to a coal liquid after dermal administration. Abstract No. 371. Toxicologist 6(1):94.

Maruoka, S., and S. Yamanaka. 1980. Production of mutagenic substances by chlorination of waters. Mutat. Res. 79:381–386.

McGill, W. B., and M. J. Rowell. 1980. Determination of oil content of oil contaminated soil. Sci. Total Environ. 14:245–253.

McGuire, M. J., and Suffet, I. H. 1983. Treatment of Water by Granular Activated Carbon. Advances in Chemistry Series 202. American Chemical Society, Washington, D.C.

Middleton, F. M., W. Grant, and A. A. Rosen. 1956. Drinking water taste and odor: Correlation with organic chemical content. Ind. Eng. Chem. 48:268–274.

Miller, F. J., D. E. Gardner, J. A. Graham, R. E. Lee, Jr., J. Bachmann, and W. E. Wilson. 1979. Particle Size Considerations for Establishing a Standard for Inhaled Particles. U.S. Environmental Protection Agency, Research Triangle Park, N.C.

Modell, M., R. P. deFilippi, and V. Krukonis. 1978. Regeneration of activated carbon with supercritical carbon dioxide. Presented before the American Chemical Society, Division of Environmental Chemistry, Miami, Fla.

Moller, M., and I. Alfheim. 1983. Mutagenicity of air samples from various combustion sources. Mutat. Res. 116:35–46.

Nguyen, T. V., J. C. Theiss, and T. S. Matney. 1982. Exposure of pharmacy personnel to mutagenic antineoplastic drugs. Cancer Res. 42:4792–4796.

Novotny, M. 1985. Analytical chromatography: The current situation and future directions, pp. 318–365. In B. L. Shapiro (ed.). New Directions in Chemical Analysis. Texas A&M University Press, College Station, Tex. (502 pp.)

NRC (National Research Council), Panel on Quality Criteria for Water Reuse. 1982. Quality Criteria for Water Reuse. National Academy Press, Washington, D.C.

Oelert, H. H. 1969. Atypical gel chromatography in the system Sephadex-isopropanol. Z. Anal. Chem. 244:91–101. (In German; English abstract.)

Ogner, G., and M. Schnitzer. 1970. The occurrence of alkanes in fulvic acid, a soil humic fraction. Geochim. Cosmochim. Acta 34:921–928.

Pederson, T. C. 1983. Biologically active nitro-PAH compounds in extracts of diesel exhaust particulate, pp. 227–245. In D. Rondia, M. Cooke, and R. K. Haroz (eds.). Mobile Source Emissions Including Polycyclic Organic Species. D. Reidel, Dordrecht, Holland.

Pederson, T. C., and J.-S. Siak. 1981. The role of nitroaromatic compounds in the direct-acting mutagenicity of diesel particle extracts. J. Appl. Toxicol. 1:54–60.

Pellizzari, E. D. 1982. Analysis for organic vapor emissions near industrial and chemical waste disposal sites. Environ. Sci. Technol. 16:781–785.

Pellizzari, E. D., and K. J. Krost. 1984. Chemical transformations during ambient air sampling for organic vapors. Anal. Chem. 56:1813–1819.

Pellizzari, E. D., J. E. Bunch, R. E. Berkley, and J. McRae. 1976. Collection and analysis of trace organic vapor pollutants in ambient atmospheres: The performance of a Tenax GC cartridge sampler for hazardous vapors. Anal. Lett. 9:45–63.

Pellizzari, E. D., L. W. Little, C. Sparacino, T. J. Hughes, L. Claxton, and M. D. Waters. 1978. Integrating microbiological and chemical testing into the screening of air samples for potential mutagenicity, pp. 331–351. In M. D. Waters, S. Nesnow, J. L. Huisingh, S. S. Sandhu, and L. Claxton (eds.). Application of Short-Term Bioassays in the Fractionation and Analysis of Complex Environmental Mixtures. Plenum, New York.

Pellizzari, E., B. Demian, and K. Krost. 1984. Sampling of organic compounds in the presence of reactive inorganic gases with Tenax GC. Anal. Chem. 56:793–798.

Penzer, G. 1980. Molecular emission spectroscopy (fluorescence and phosphorescence), pp. 70–114. In S. B. Brown (ed.). An Introduction to Spectroscopy for Biochemists. Academic Press, New York.

Preidecker, B. L. B. 1980. Bacterial mutagenicity of particulates from Houston air. Environ. Mutagen. 2:75–83.

Randall, L. G. 1984. Carbon dioxide based supercritical fluid chromatography: Column efficiencies and mobile phase solvent power, pp. 135–169. In S. Ahuja (ed.). Ultrahigh Resolution Chromatography. ACS Symposium Series 250. American Chemical Society, Washington, D.C.

Rappaport, S. M., Y. Y. Wang, E. T. Wei, R. Sawyer, B. E. Watkins, and H. Rapoport. 1980. Isolation and identification of a direct-acting mutagen in diesel-exhaust particulates. Environ. Sci. Technol. 14:1505–1509.

Sato, T., T. Momma, Y. Ose, T. Ishikawa, and K. Kato. 1983. Mutagenicity of Nagara River sediment. Mutat. Res. 118:257–267.

Schoeny, R. 1985. Exploratory Research on Mutagenic Activity of Coal-Related Materials Using Statistical Evaluation. DOE/PC/62999-6. Kettering Lab, Cincinnati, Ohio. (Available from NTIS as DE85013667/XAB.) (19 pp.)

Schoeny, R., D. Warshawsky, and G. Moore. 1986. Non-additive mutagenic responses by components of coal-derived materials. Am. Chem. Soc., Div. Fuels Chem. Preprints 31(2):147–155.

Schoeny, R., D. Warshawsky, L. Hollingsworth, M. Hund, and G. Moore. 1980. Mutagenicity of coal gasification and liquefaction products, pp. 461–475. In M. D. Waters, S. S. Sandhu, J. L. Huisingh, L. Claxton, and S. Nesnow (eds.). Short-Term Bioassays in the Analysis of Complex Environmental Mixtures II. Plenum, New York.

Selby, C., J. Calkins, and H. Enoch. 1983. Detection of photomutagens in natural and synthetic fuels. Mutat. Res. 124:53–60.

Shapiro, J. 1961. Freezing-out, a safe technique for concentration of dilute solutions. Science 133:2063–2064.

Silverstein, R. M., and G. C. Bassler. 1963. Spectrometric Identification of Organic Compounds. John Wiley & Sons, New York (377 pp.)

Smith, R. D. 1983. New approaches combining chromatography and mass spectrometry for synfuel analysis, pp. 332–352. In C. W. Wright, W. C. Weimer, and W. D. Felix (eds.). Advanced Techniques in Synthetic Fuels Analysis. PNL-SA-11552. CONF-811160. U.S. Department of Energy, Technical Information Center, Oak Ridge, Tenn. (Available from NTIS as DE83015528.)

Smith, R. D., and H. R. Udseth. 1983. Mass spectrometry with direct fluid supercritical injection. Anal. Chem. 55:2266–2272.

Smith, R. D., H. R. Udseth, and H. T. Kalinoski. 1984. Capillary supercritical fluid chromatography/ mass spectrometry with electron impact ionization. Anal. Chem. 56:2971–2973.

Springer, D. L., M. L. Clark, D. H. Willard, and D. D. Mahlum. 1982. Generation and delivery of coal liquid aerosols for inhalation studies. Am. Ind. Hyg. Assoc. J. 43:486–491.

Springer, D. L., R. A. Miller, W. C. Weimer, H. A. Ragan, and R. L. Buschbom. 1984. Effects of Inhalation Exposure to SRC-II Heavy and Middle Distillates. PNL-5273. Battelle Pacific Northwest Labs, Richland, Wash. (Available from NTIS as DE85004327/XAB.) (68 pp.)

Stamoudis, V. C. 1982. A gas chromatographic scheme, based on relative retention indices for rapid quantitative and qualitative analysis. Abstract No. 72 in Pittsburgh Conference and Exposition on Analytical Chemistry and Applied Spectroscopy, March 8–13, 1982, Atlantic City, N.J. Abstracts (unpublished).

Stamoudis, V. C., and J. C. Demirgian. 1985. Computer-Assisted Analysis of Energy-Related Complex Mixtures by Retention-Index Gas Chromatography. ANL/SER-5. Argonne National Lab, Argonne, Ill. (Available from NTIS as DE85013702/XAB.) (55 pp.)

Stamoudis, V. C., D. A. Haugen, M. J. Peak, and K. E. Wilzbach. 1983. Biodirected chemical characterization of synfuel materials, pp. 202–214. In C. W. Wright, W. C. Weimer, and W. D. Felix (eds.). Advanced Techniques in Synthetic Fuels Analysis. PNL-SA-11552. CONF-811160. U.S. Department of Energy, Technical Information Center, Oak Ridge, Tenn. (Available from NTIS as DE83015528.)

Strommen, D. P., and K. Nakamoto. 1984. Laboratory Raman Spectroscopy. John Wiley & Sons, New York (138 pp.)

Swain, A. P., J. E. Cooper, and R. L. Stedman. 1969. Large-scale fractionation of cigarette smoke condensate for chemical and biologic investigations. Cancer Res. 29:579–583.

Talcott, R., and W. Harger. 1980. Airborne mutagens extracted from particles of respirable size. Mutat. Res. 79:177–180.

Teranishi, K., K. Hamada, and H. Watanabe. 1978. Mutagenicity in *Salmonella typhimurium* mutants of the benzene-soluble organic matter derived from air-borne particulate matter and its five fractions. Mutat. Res. 56:273–280.

Tokiwa, H., K. Morita, H. Takeyoshi, K. Takahashi, and Y. Ohnishi. 1977. Detection of mutagenic activity in particulate air pollutants. Mutat. Res. 48:237–248.

Toste, A. P., D. S. Sklarew, and R. A. Pelroy. 1982. Partition chromatography—high-performance liquid chromatography facilitates the organic analysis of biotesting of synfuels. J. Chromatogr. 249:267–282.

U.S. EPA (Environmental Protection Agency). 1971. National Primary and Secondary Ambient Air Quality Standards. Appendix B: Reference method for the determination of suspended particulates in the atmosphere (high volume method). Federal Register 36:8191–8193.

U.S. EPA. 1982. Test Methods for Evaluating Solid Waste: Physical/Chemical Methods. 2nd ed. SW846. U.S. Environmental Protection Agency, Office of Solid Waste, Washington, D.C. (Available from NTIS as PB 87-120291.)

U.S. EPA. 1984. Calculation of Precision, Bias, and Method Detection Limit for Chemical and Physical Measurements. EPA 600/4-85-058. U.S. Environmental Protection Agency, Office of Research and Development, Office of Monitoring Systems and Quality Assurance, Quality Assurance Management and Special Studies Staff, Washington, D.C.

U.S. EPA. 1985. Guidelines for Preparing Environmental and Waste Samples for Mutagenicity (Ames) Testing. Environmental Monitoring System Laboratory, Las Vegas, Nev. (Available from NTIS as PB 86-120144.) (255 pp.)

van Hoof, F., and J. Verheyden. 1981. Mutagenic activity in the River Meuse in Belgium. Sci. Total Environ. 20:15–22.

Vithayathil, A. J., B. Commoner, S. Nair, and P. Madyastha. 1978. Isolation of mutagens from bacterial nutrients containing beef extract. J. Toxicol. Environ. Health 4:189–202.

Watson, J. T., and Throck. 1985. Introduction to Mass Spectrometry. Raven Press, New York. (351 pp.)

Weimer, W. C., and B. W. Wilson. 1983. Integration of chemical analysis and biological testing in the study of coal liquid toxicology. (Abstract.) Toxicol. Lett. 18(Suppl. 1):78.

Wilk, M., J. Rochlitz, and H. Bende. 1966. Säulenchromatographie von polycyclischen aromatischen Kohlenwasserstoffen an lipophilem Sephadex LH-20. J. Chromatogr. 24:414–416.

Wilson, B. W., R. Pelroy, and J. T. Cresto. 1980. Identification of primary aromatic amines in mutagenically active subfractions from coal liquefaction materials. Mutat. Res. 79:193–202.

Wilson, B. W., A. P. Toste, R. A. Pelroy, B. Vieux, and D. Wood. 1981. Accurate Mass/Metastable Ion Analysis of Higher-Molecular-Weight Nitrogen Compounds in Coal Liquids. CONF-801039-7. Battelle Pacific Northwest Labs, Richland, Wash. (Available from NTIS as PNL-SA-8852.) (21 pp.)

Wood, E. J. 1980. Atomic absorption spectroscopy, pp. 320–335. In S. B. Brown (ed.). An Introduction to Spectroscopy for Biochemists. Academic Press, New York.

Wright, B. W., R. D. Smith, and H. R. Udseth. 1984. Approaches and applications of supercritical fluid chromatography and supercritical fluid chromatography-mass spectrometry techniques. Abstract No. 594 in The Pittsburgh Conference and Exposition, March 5–9, 1984, Atlantic City, N.J. Abstracts (unpublished).

Wright, B. W., E. K. Chess, H. T. Kalinoski, R. D. Smith, and C. W. Wright. 1985. Comparative analysis of nitro-PAH by capillary gas and supercritical fluid chromatography methods. Abstract No. ANYL 8 in American Chemical Society, Abstracts of Papers, 189th ACS National Meeting, Miami Beach, Fla., April 28–May 3, 1985. American Chemical Society, Washington, D.C.

Wright, C. W. 1984. Comparative analysis of four quantitative methods for coal liquids analysis using capillary column chromatography. Abstract No. 831K in The Pittsburgh Conference and Exposition, March 5–9, 1984, Atlantic City, N.J. Abstracts (unpublished).

Wright, C.W., and D. D. Dauble. 1986. Effects of Coal Rank on the Chemical Composition and Toxicological Activity of Coal Liquefaction Materials. PNL-5805. U.S. Department of Energy, Washington, D.C. (Available from NTIS as DE86011015.) (67 pp.)

Wynder, E. L., and D. Hoffmann. 1965. Some laboratory and epidemiological aspects of air pollution carcinogenesis. J. Air Pollut. Control Assoc. 15:155–159.

Wynder, E. L., and G. Wright. 1957. A study of tobacco carcinogenesis. I. The primary fractions. Cancer 10:255–271.

5

Interpretation and Modeling of Toxicity-Test Results

INTRODUCTION

Nearly all the problems inherent in the toxicity assessment of single agents—from the selection and measurement of health end points through the specification of dose or exposure and the details of experimental design to the prediction of risk—are exacerbated in the toxicity assessment of complex chemical mixtures. Each of these problems is amplified by the possibility of interactions or other unpredicted behavior among the mixture components.

The methods of biostatistics and quantitative modeling play important roles in toxicology. The statistical/mathematical methods may be applied in four ways. First, they provide some guidance in developing an overall strategy to test the toxicity of mixtures economically. Second, they offer directions for experimental design so that maximal information can be extracted from the results of studies that cannot address directly every possible combination, dose, and regimen of exposures. Third, they offer a framework for interpreting data that often present unusual features and questions. Fourth, they offer a choice of approaches for estimating the consequences of biologic assumptions about interactions with an eye to developing more realistic models.

Quantitative models have been applied to assess the joint toxicity or effect of chemicals since the seminal work of Bliss (1939), Plackett and Hewlett (1948), and Finney (1952). This work has not focused on the unintentional, usually low-level exposure to chemicals in the environment, but was undertaken to understand whether the joint application of two pesticides or the joint administration of two or more drugs led to outcomes that were greater than the sum of the outcomes predicted on the basis of individual applications of the substances.

The environment exposes us to thousands of chemicals, some appearing intentionally (such as food additives) and some appearing as adventitious contaminants. Even if humans were exposed to no more than 100 potentially toxic agents, the possibility of unusual or unexpected combined effects is sizable. The matrix of only single-dose combinations of two of these agents at a time would contain 4,950 cells. If the probability that the presence of one agent influences the toxicity of another were as low as 0.01, there would still be 50 combinations of two agents that would interact in some unpredicted way.

In Chapter 1, "interaction" was referred to as deviation from dose-additive behavior expected on the basis of dose-response curves obtained with individual agents. To a biostatistician, the definition includes information on the underlying dose-response model and on units of measure. For ordinary linear models, "interaction" refers to a departure from response additivity. Different measures of response can lead to different conclusions concerning departure from additivity. For nonlinear models, such as log-logistic, log-linear, log-probit, and multistage models, "interaction" can be defined as a departure from additivity for a transformation of the response variable. For example, with multiplicative models, there is additivity for the logarithm of response. Thus, when the word "interaction" is used, one must make certain of the units in which toxicity or dose is expressed, as well as the assumptions about the nature of joint action that was predicted by the model used. Kodell and Pounds (in press) reviewed the use of the nomenclature and noted that different scientific disciplines define these concepts differently; moreover, even within a discipline, there may be disagreement about the definition of specific concepts.

Models, the major topic of this chapter, constitute attempts to quantitatively describe the important, measurable aspects of the complex biology or toxicology of chemical mixtures. To date, all models have been imperfect, yet they have been helpful in estimating the hazard associated with exposure to a complex mixture. Most toxicology experiments obtain fairly limited data, because resources are constrained. Models are a means of characterizing and summarizing experimental results, often relating them to underlying biology. If this is done well, it becomes possible to extend limited experimental results beyond the dose range from which they were developed. Models furnish a way to test the consistency of experimental results with biologic assumptions or hypotheses. Tests for the existence of interaction are based on model assumptions.

Estimation of responses to low-level exposures introduces a difficult problem in assessing the toxicity of mixtures, as well as of single agents. Experiments are usually conducted at exposures far greater than those occurring environmentally. The logical problem in extending these results lies in the fact that the presence (or absence) of interaction at high doses does not guarantee its presence (or absence) at other doses. Hence, experimental results alone are likely to be inadequate for assessing the toxicity of a mixture or the presence of substantial interaction at doses comparable with human exposure. Models have been constructed to estimate responses that are not experimentally observable.

Models supply a framework for guiding experimentation or for experimental designs that might be quite different from those traditionally used. For example, where there is sufficient confidence in a model for regulatory use, it might not be necessary to conduct experiments at joint doses; reliance on experimental results with separate components might be sufficient.

Models contribute to the estimation of the toxicity of complex mixtures. One important facet of a complex-mixture research strategy is the development and improvement of models that relate toxic response to exposure. The development of models requires greater attempts to understand and describe biologic mechanisms, so that better mathematical models can be developed.

What to Test? Whole Mixtures versus Separate Components of Mixtures

Approaches to estimating the toxicity of mixtures can emphasize the testing of the mixtures themselves or the testing of their individual components, eventually building up to the mixture (see Chapter 3). When separate components are tested, the toxicity of a mixture is estimated by testing more and more complex combinations of the components or by applying models to estimate the toxicity of the combination of constituents. When whole mixtures are tested, the components of the mixture and their relative quantities might be unknown, and the mixture can at times be divided into other mixtures that can be examined for their contribution to overall toxicity.

The choice of approach depends on the information available, the complexity of the mixture, and the goals of the assessment. If the objective of testing is the toxicity assessment of the emissions from a dump site with a small number of known constituents, the approach emphasizing knowledge about the constituents, individually and jointly, might be useful in estimating the potential toxicity of the dump site at defined exposures. This approach could also be used to estimate the toxicity of various combinations of a limited number of the constituents. If the objective is the toxicity assessment of a very complex mixture, such as gasoline, the constituents might be too numerous (or even unknown) to permit testing combinations and all components. Sometimes, a combination of the two approaches is optimal. For example, if the objective is the identification of key toxic components or fractions in a mixture, various fractions of the mixture could be tested to identify the most toxic ones. Separate components of those fractions, in turn, could be tested in combination, to identify interactions among them.

This chapter emphasizes relatively simple mixtures (usually of only two materials) and the use of models to predict responses to their joint exposure, given information about their constituents. This emphasis eases the development of concepts, models, and experimental designs. Many of the results can be extended to more complex mixtures whose components or major fractions can be defined. The models developed to date are largely for the estimation of cancer risks associated with lifetime exposure.

This chapter is organized in four sections. The first discusses a number of mathematical dose-response models, how they can be used to evaluate the experimental strategies presented in Chapter 3, and how to apply such strategies. The second section addresses the importance of knowledge of pharmacokinetics in mixture toxicity assessment. The third section provides experimental design criteria and guidelines that arise from the previous considerations regarding dose-response models. Finally, the fourth section offers recommendations for research that could generate better methods of dose-response modeling and interpretation. Appendixes providing the mathematical bases for the assertions made are included at the end of the report.

Implications for Strategies

The most difficult issue posed by an assessment of a complex mixture's toxicity is the possibility of unexpected interactions among the components of the mixture. Complex-mixture toxicology is a young science and depends heavily on existing tools and concepts. This chapter, while discussing directions for development, emphasizes currently available tools and provides some guidance for their use.

Evaluation of Mixture Components

The specific experimental strategy selected depends, in part, on how much is known about the mechanisms leading to a particular biologic response. Carcinogenesis models provide an example of the utility and limitations of models. Appendix E illustrates the use of these models for mixtures.

Expectation of Additivity at Low (Response) Doses

The multistage model is one of the most widely used models to predict the incidence of cancer at low exposures from incidence data associated with higher doses. It describes the dose-response relationship over a full range of doses and thus permits response extrapolation at untested doses. At sufficiently low doses, the excess response or risk associated with exposure to a mixture of two components will be equal to the sum of the excess responses (risks) associated with each of the components; that is, low-dose additivity will hold. "Low" dose as used in Appendix E is defined as a dose associated with an excess risk of less than 0.1–1% and a small relative risk of cancer in the ex-

TABLE 5-1 Incremental Probability of Cancer Due to Exposure to Carcinogenic Agents X and Y

	Cancer Probability at Agent X Dose		
Agent Y Dose	0	*A* (low)	*B* (high)
0	0	1×10^{-5}	9.0494×10^{-2}
A' (low)	1×10^{-5}	2.0045×10^{-5}	9.0892×10^{-2}
B' (high)	9.0494×10^{-2}	9.0892×10^{-2}	9.8403×10^{-1}

posed group (e.g., less than 1.01).* The known human carcinogens with environmental exposures at "high" doses for the general population include cigarette smoke, ultraviolet light, and radon. Some workplace exposures might also lead to high risk for some populations.

Appendix E gives an example of a mixture with a much greater than additive effect in a specific multistage model; in fact, the example assumes one of the largest synergistic effects that could be practically estimated with a two-dose bioassay design with 50 animals in each of four treatment groups (control for each agent; control agent X, treatment agent Y; control agent Y, treatment agent X; treatment agents X and Y). Table 5-1 summarizes the results of the example. At dose *A* of agent X and no exposure to agent Y, the probability of cancer is 1×10^{-5}; a similar probability exists for dose *A'* of agent Y and no exposure to agent X. Given simultaneous exposure to agents X and Y at doses *A* and *A'*, the risk is 2.0045×10^{-5}, compared with an expectation of 2.000×10^{-5} if the risk were additive. If exposure to one of the agents, say X, increases to dose *B*, which would give a response of 9.0494×10^{-2}, and exposure to the other agent remains at dose *A'*, the risk is 9.0892×10^{-2}, compared with the additivity assumption estimate of 9.04941×10^{-2}. Only when both doses are relatively "high," at *B* and *B'*, both giving an individual response of 9.0494×10^{-2}, does the additivity assumption fail. The estimate from joint exposure is 9.8403×10^{-1}, compared with an additivity-only estimate of 1.81784×10^{-1}.

Several assumptions are inherent in these results. Appendix E formally states the assumptions, and several of them are examined here. First, the doses considered are assumed to be those at the target organ or directly proportional to those at the target organ. Second, the relative size of the background tumor rate influences the accuracy of the additivity result. For the multistage model, the computed additivity result depends on the assumption that the relative risks of each agent are small (e.g., less than 1.01). The additivity property can also be shown to hold for mixtures of three or more components, as well as for other

*The multistage model considered in the appendix also assumes that the components affect single and different transition rates between stages in the model.

dose-response models, such as the Moolgavkar-Knudson, linear-logistic, and multiplicative models. The difficulty in invoking this result, however, can be the ignorance about how low doses must be before additivity is achieved.

One important result of the development presented in Appendix E is that, for a very wide range of models, if the model can be specified, the toxicity of the mixture can be estimated through knowledge of the toxicity of the mixture components. This may be true even where component-wise additivity does not hold.

Appendix E presents a multistage model for a two-material mixture. Estimation of the values of parameters for this conceptually simple model can require considerable data—including toxicity data on combinations of the two mixture components at several dose combinations. For more complex mixtures, estimation of values of model parameters might not be practical. The problem can be simplified if another model, such as a logistic model, is used to describe the dose-response surface.

Models cannot be fully validated, so their utility depends on the confidence that one places in them, which in turn is related to the reasonableness of their results and their concordance with biologic knowledge and experimental results. Even in the case of carcinogenesis, for which there has been much work on models, some assumptions might not be met for specific materials. For example, the usual application of the multistage model ignores potential actions of a promoter. If a chemical were identified as a likely promoter, an alternative approach explicitly treating promoters would be required. Some specifications of the multistage or Moolgavkar-Knudson model might be applied to address this issue.

Appendix F reviews the use of models for developmental toxicology. Given the multiplicity of end points and mechanisms, no single model is likely to be appropriate. Further research is needed before any of these models is likely to be accepted.

Appendix D develops and illustrates how new models might be developed. The set of dose-response models applied to biologic phenomena is small and consists of the linear, multiplicative, linear-logistic, multistage, Weibull, and probit models and a few others. These have been used to describe many biologic phenomena. It has been suggested that using a set (or perhaps all) of them to provide risk estimates might be appropriate for situations where no single model truly reflects the biology. The set (or some subset) of models could be applied simultaneously to define a range or limits of the toxic risk for a mixture at a given dose. One concern with this approach, however, is that all the models used might be inappropriate. The resulting range might then not bracket the true risk. A further concern is that one or more of the models in a set might be so poor in describing the biology that they could yield results far from the true risks—and thus make the range of results so broad as to be essentially meaningless.

The toxicity of individual components and combinations can be estimated experimentally through the use of generalized linear models. Appendix G has a description of this approach and an application of some of the fire-toxicology data described in the Appendix C case studies. These methods are relatively easy to apply, require few assumptions, and use currently available computer software. The principal drawback is that many data points are required, so experiments must incorporate far more doses than is usually done or can usually be done. In addition, it can be dangerous to extrapolate outside the range of observed doses with these methods, because interactions between components might be dose-dependent in some way that does not clearly manifest itself in the observed data.

Evaluation of Composite Mixture

A mixture can be treated as a single unknown substance and its toxicity evaluated accordingly. Most of the problems associated with this approach resemble those which arise in the toxicology of single substances and are addressed by similar models. The major questions are as follows:

- How big a departure is needed from specific mixtures before toxicity changes?
- What happens when a component is added or removed?

The models and statistical methods available to address these issues were developed for single substances; although the concepts are the same, a routine application of these models to complex mixtures could cause misleading interpretations of results. For example, misinterpretation could arise if the relative composition of the mixture changed as a result of chemical interactions or as a result of variations in differential absorption, distribution, metabolism, or elimination with dose or some other factor independent of the mixture. Current pharmacokinetic models attempt to describe the fate of a single substance and ignore possible complexities introduced by a mixture. Such complexities could distort the predicted concentration and composition of a mixture at the target sites where toxic effects are initiated. Pharmacokinetic assumptions about compositional changes in the mixture would also have to apply to all the animal species involved in any extrapolation. If the composition of the mixture changed between administration and delivery to the target organ in the rat, similar changes would have to occur in humans if an extrapolation were to be valid. In addition, assumptions need to be made about the uniformity of responses of individuals within a species.

The dose-extrapolation issue is also complicated by mixtures. Models are used to help to predict toxicity of low ambient exposures from observed toxicity of high experimental exposures. According to the most widely used carcinogenesis models, additivity of effects is a reasonable assumption if all the exposures to all the toxic components of a mixture are small. For high doses, as noted earlier, substantial deviations from additivity can be observed. If a mixture were tested at a high dose, the estimated toxicity would include the contribution of interactions; that is, the estimated toxicity of the mixture would be the sum of the toxicities of the individual components plus the interaction terms (which could be negative if there were antagonism or inactivation of some components). Extrapolation of toxicity to "low" doses with models developed for a single toxic agent would reflect the presence of the interaction terms, although those terms would not have been explicitly measured.

For a dose-extrapolation model to be satisfactory, when the doses of the components are sufficiently small so that the incidence of the toxicity of concern associated with each is below about 0.1–1% and the relative risks are small (e.g., less than 1.01), the interaction terms should approach zero. If explicit allowance for interactions is not made in the extrapolation model, the toxicity of the mixture at low doses could be incorrectly estimated. If an interaction term were positive, the extrapolated risk of the mixture could be overestimated at low doses; if it were negative, the risk could be underestimated.

With the strategy in which toxicity information is collected only for the mixture itself, no information on the toxicity of individual components would be collected and it would be impossible to account for interaction explicitly in the analysis indicated by the dose-extrapolation model. Practical considerations, such as ignorance about the components or the existence of too many components, might require the assessment of the composite as though it were a single substance. In such a situation, concern about bias will be greatest when the toxicity due to interaction is relatively large, compared with the toxicity of the individual constituents.

A compromise for this situation would be to divide the mixture into individual components or fractions. It might be possible to define fractions so as to minimize or maximize interactions within each fraction. The goal involved for the specific test will drive the fractionation carried out.

COMPARATIVE EVALUATION

This approach is based on the premise that studies of the toxicity of one mixture will provide information about the toxicity of a related mixture. A prototype for examining a body of such data is the set of cross-mixture extrapolation methods applied by Harris (1983a,b) and based on the methods of DuMouchel and Harris (1983). Recent work on these methods has also been done by Verducci (1985) and Laird and Louis (1986).

Harris illustrated the approach to estimate the lung-cancer risks of diesel emissions, given estimates of these risks for two similar substances, coke-oven emissions and roofing-tar emissions. He does this by assuming that the results of laboratory bioassays of all three substances can be used to estimate the relative carcinogenic potencies of diesel emissions and coke-oven or roofing-tar emissions. These potency estimates are then applied to the lung-cancer risk estimates of the latter two substances, to allow an estimate of lung-cancer risks associated with diesel emissions. Fundamental to this approach is the assumption that the relative carcinogenic potencies of diesel emissions are the same for humans and for the test systems used in the analysis. Harris acknowledged that that assumption is, at best, an approximation.

The reasonableness of the approximation is unknown. The Harris example yielded similar upper confidence limits for excess human lung-cancer risks. Considerable additional experience with this approach is needed for its value to be established. The advantage of the approach is that it makes use of existing data, including those from short-term studies. However, the uncertainty in the approach makes it, at best, useful for screening, in which it might indicate whether a given mixture is considerably more or less toxic than a similar mixture on which more information is available.

All the above methods could be included in a general strategy defined as a "global approach" to assess the toxicity of a mixture. The strategy confines itself to a class of mixtures, such as gasoline, and provides a way to estimate the toxicity of mixtures in the class, taking into account the differences in composition of the mixtures. The strategy also avoids the need to assess the toxicity of every mixture in the class.

The definition of the strategy will vary with the class of mixtures and the extent of information available. In general, however, the global perspective advocates testing extreme formulations of a mixture, as well as the more commonly used formulations. Hence, information would be available on the greatest possible variation in the mixture formulations under study. In addition, the approach recommends the testing of relatively homogeneous fractions or components of the mixture individually or in combination with a few other substances.

Satisfying that approach requires new data on different mixture formulations and on mixture components and their combination. Application of some existing statistical theory in experimental design can reduce the data to be collected. Fractional factorial designs are particularly useful. They permit estimation of the individual effects of each factor, and they provide information on the most important interactions among factors.

The use of fractional factorial designs depends on the assumption that non-high-order interactions are of consequence. Looking only for low-order interactions and main effects permits designs that place a few animals in each of a number of previously defined dose groups. For example, suppose we wish to study simultaneously the effects of 15 constituents, each at a high concentration and a low concentration. There are 2^{15} possible combinations of the 15 materials, each of which can be at one of two levels. There are, however, fractional factorial designs that involve only 32 (2^5) possible combinations and permit estimation of all the effects of the 15 individual constituents and interactions among all pairs of any specified subset of six of the constituents (John, 1971). However, even 32 lifetime full-scale animal bioassays might be economically impossible. A way of incorporating short-term tests needs to be developed (see Chapter 3). Alternatively, models, such as the multistage model, might be applied where appropriate and where additivity of response could be assumed in place of assessment of some combination of factors.

The comprehensiveness and flexibility of the global approach make it an appealing framework for addressing complex-mixture toxicity. The specific methods within the approach for assessing the collected and available information require better definition. The best judgments about the promise of this approach, however, will arise from its application to specific mixture problems.

PHARMACOKINETIC MODELS

Pharmacokinetic models can be used to describe the absorption, distribution, metabolism, binding, and elimination of a substance that enters the body. They thereby provide useful information on the formation and distribution of the reactive metabolites responsible for the induction of toxic effects in individual tissues. These models can be used to describe the relationship between the dose administered in toxicity tests and the dose delivered to the target tissue.

Pharmacokinetic models are mathematical/physiologic constructs that envisage a biologic system made up of a smaller or larger number of relevant physiologic compartments. The interactions among these compartments are often described by a set of differential equations (Gibaldi and Perrier, 1975) and may be developed from more general information based on the anatomy and physiology of the test animal, binding and metabolism in specific organs, and the solubility in various organs and tissues of the chemical under test (Ramsey and Andersen, 1984; Andersen et al., 1987).

One of the potential applications of pharmacokinetics to risk assessment involves the use of pharmacokinetic models to determine the effective dose of a compound of interest that reaches the target tissue. Knowing the dose delivered to the target tissue should then lead to more accurate estimates of risk (Hoel et al., 1983; Whittemore et al., 1986).

The internal tissue dose for routes of exposure other than the ones used in the experimental situation may be developed from pharmacokinetic models whenever measurements of the physiologic and biochemical constants associated with the several possible routes are available (Chen and Blancato, 1987). Extrapolation between species is facilitated when physiologic models can be used to predict the delivered dose in the species of interest (Andersen et al., 1987). When not all the relevant model characteristics can be measured directly in humans, however, estimates of values of the parameters of the equations applying to humans must be obtained by scaling the corresponding animal values.

The application of pharmacokinetic models to complex mixtures is more involved than their application to single chemicals. The simplest situation with a mixture would occur if there were no saturation or capacity restrictions on any of the activation, deactivation, or tissue-binding processes and no chemical interactions among mixture components. Pharmacokinetic studies could be carried out on the components of the mixture, and the concentration of a particular substance or metabolite in each organ could be taken to be the sum of the tissue doses contributed by each of the components of the mixture. One way to determine whether this simple situation occurs is to conduct pharmacokinetic studies on the distribution, metabolism, and clearance of each of the components separately and then for the mixture as a whole. The extent to which the mixture results can be predicted by combining the results for the individual components indicates the extent of interaction among the components.

If the pharmacokinetic behavior of the mixture differs from that predicted on the basis of results obtained for the individual components, one possibility is that the same components and metabolites occur in the mixture study and in the component studies, but in different proportions. That would suggest competition among the components for activation, deactivation, and elimination through the same pathways. Another possibility is that products formed from the mixture are different from those formed from the components. That could suggest chemical interactions among the components, rather than competition with common pathways. In either case, the pharmacokinetic behavior of the mixture could not be predicted solely from the pharmacokinetic behavior of the components. The components would need to be considered jointly for modeling of the saturation of the resources for which they compete and modeling of the joint reaction products.

Whether the components of a complex mixture will interact is often unknown, and the interaction problem usually has to be approached empirically. In general, a completely empirical approach to identifying and measuring interactions is impossible. However, some interactions can be assumed to occur only among a few components; fractional factorial designs may be used to test for interactions with a relatively small number of test combinations.

In developing pharmacokinetic models for toxicity assessment of mixtures, consideration should be given to the following questions:

- Which constituents (or transformation products derived therefrom) are known to be associated with toxicity and therefore need to be modeled?
- Can pharmacokinetic data on individual components be combined to predict the pharmacokinetic behavior of the mixture, or do some components alter the pharmacokinetic behavior of others?

- Do capacity restrictions or saturable phenomena preclude simple addition of tissue concentrations of individual components to estimate the tissue concentration of the mixture?
- Do interactions among the components produce new toxicants that are absent when the components are evaluated individually?
- Do the nature and extent of interactions among mixture components differ among species involved in extrapolation?
- What is known about the mechanisms of toxicity of the mixture and of the individual components?

EXPERIMENTAL DESIGN

Experimental design that seeks the optimal allocation of resources is an important part of the testing of potentially toxic materials. The primary goals of such tests are the identification of toxic agents, the establishment of dose-response relationships, and the estimation of risks at environmental levels of exposure. Another possible objective is to provide information on the mechanism by which two toxicants can interact, as in studies on initiation and promotion systems of carcinogens. These different goals lead to different experimental designs that involve different numbers of dose groups, dose levels, and numbers of animals at the chosen doses. Different materials can be administered simultaneously for assessing the toxicity of the mixture as a whole, separately for evaluating the effects of the mixture components, or sequentially for studying initiation/promotion.

This section raises some issues that warrant consideration in designing experiments with mixtures. Work in this field needs to be extended considerably and tested with real mixtures before guidelines can be developed. Hence, the purpose of this section is to discuss, in general terms, issues and approaches that might be relevant to the design of future experiments that assess the toxicity of mixtures, rather than to provide hard and fast design instructions.

Earlier in this chapter it was noted that the presence of interactions at high doses need not imply the presence of appreciable interaction at low doses. Hence, experiments carried out at several doses were encouraged. The precise objectives of the research and the nature of the assumptions made will help to determine the number of appropriate doses, as well as other aspects of experimental design. Thus, the discussion of the impacts of some of the possible objectives and assumptions on the experimental design are presented here to motivate further thought on specific mixture studies.

Clearly, we have not reached the point of integrating the many dimensions of experimental design for mixtures to establish firm recommendations for practical use. Much of the work discussed below concentrates on binary mixtures, but does provide some useful insight on experimental design. Results here, as elsewhere, depend on the objectives of the study at hand, as well as on the underlying model assumptions. An important future task is the development of detailed designs for assessing the toxicity of complex mixtures by collecting information about mixture constituents, both singly and in combination.

IDENTIFICATION OF TOXIC AGENTS

Studies conducted for the identification of hazardous chemicals are often labeled screening experiments. From these experiments, only a qualitative yes-no answer is required as to the presence of carcinogenic effects. The obvious approach for a screening experiment is to space doses and allocate animals in such a way as to maximize the likelihood of discovering that a material has toxic potential. That implies exposure at some high dose—high enough to elicit a statistically significant response when treated animals are compared with control animals. Thus, in testing for the effects of a single chemical, the optimal design involves only two treatment groups (a control group and a high-dose group) chosen so as to maximize the difference between the toxic-response rates in the two groups.

Consider now the case of a binary mixture. In general, a factorial design for two chemicals, C_1 and C_2, will involve $(k_1 + 1) \times (k_2 + 1)$ dose combinations, where k_1 and k_2 are the numbers of nonzero doses of C_1 and C_2, respectively, shown in Table 5-2.

Determination of an experimental design requires specification of the numbers of doses, k_1 and k_2, of C_1 and C_2, the actual dose pairings (d_{1i}, d_{2j}) to be used, and the number of animals, n_{ij}, to be allocated to each treatment combination $(i = 0, 1, \ldots, k_1; j = 0, 1, \ldots, k_2)$.

In what follows, we will consider a minimal factor design consisting of the four dose combinations $(d_{1i}, d_{2j}) = (0, 0), (d_{11}, 0), (0, d_{21})$, and (d_{11}, d_{21}). Scaling the doses so that $d_{11} = d_{21} = 1$, it will be convenient to refer to them as the (0, 0), (1, 0), (0, 1), and (1, 1) cells of this four-point design. In conjunction with the (0, 0) cell, the (1, 0) and (0, 1) cells provide for an evaluation of the

TABLE 5-2 A Factorial Experimental Design for Binary Mixtures Involving Two Chemicals C_1 and C_2

	C_2				
	d_{20}	d_{21}	. . .	d_{2k_2}	Total
C_1					
d_{10}	n_{00}	n_{01}	. . .	n_{0k_2}	$n_{0\cdot}$
d_{11}	n_{10}	n_{11}	. . .	n_{1k_2}	$n_{1\cdot}$
.					
.					
.					
d_{1k_1}	n_{k_10}	n_{k_11}	. . .	$n_{k_1k_2}$	$n_{k_1\cdot}$
Total	$n_{\cdot 0}$	$n_{\cdot 1}$	. . .	$n_{\cdot k_2}$	n

toxic effects of C_1 and C_2 separately and with joint exposure (1,1) cell further providing for an assessment of interaction.

For purposes of detecting any toxic effects attributable to C_1 and C_2, the doses d_{11} and d_{21} should, as argued previously, be selected so as to maximize the probability of inducing an adverse response based on the information provided by the (1,0) and (0,1) cells. However, this practice might result in overly severe effects in the (1,1) cell in which both C_1 and C_2 are administered simultaneously. In carcinogen bioassays with single compounds, for example, it is common practice to choose the doses as large as possible without appreciably reducing survival or growth rates, other than as a result of tumor occurrence (National Toxicology Program, 1984). With a binary mixture, the use of such maximal tolerated doses (MTDs) might result in markedly reduced survival or growth when compounds are administered together at their respective MTDs.

Wahrendorf et al. (1981) considered optimal four-point designs for testing for interaction at high doses between C_1 and C_2, assuming the log-linear model,

$$\log (1 - p_{ij}) = \mu + \alpha_i + \beta_j + \delta_{ij}, \tag{5-1}$$

where p_{ij} is the probability of a toxic response in the $(i,j)^{\text{th}}$ cell $(i,j = 0,1)$. Here, μ represents a measure of the spontaneous-response rate; α_1 and β_1, the effects of C_1 and C_2 separately; and δ_{11}, the interaction between C_1 and C_2. (The remaining parameters α_0, β_0, δ_{00}, δ_{10}, and δ_{01} are all zero.)

By minimizing the variance of the maximum-likelihood estimator of δ_{11} in the null case of no interaction, they found the optimal allocation of animals to the (0,0), (1,0), (0,1), and (1,1) cells in the design. In general, they demonstrated that, for the case of carcinogens that were expected to increase background rates by a factor of 2–4 when operating singly at the doses administered, the optimal allocation of subjects needed to determine whether the combination of the two materials yields an interactive effect is roughly as follows:

Controls (0,0):	0.12
Each material alone (1,0 and 0,1):	0.25
Two materials in combination (1,1):	0.38

These proportions will vary with background response rate, the toxicities of the two materials, and the magnitude of any presence of interactive effects. In the examples considered by Wahrendorf et al. (1981), the use of these unbalanced optimal designs can provide gains in efficiency of approximately 20%, compared with the balanced design in which equal numbers of subjects are assigned to the four treatment groups.

It is important to recognize that the optimal allocations given above for the log-linear model in Equation 5-1 might not apply when different assumptions are made. For example, using the same methods as used by Wahrendorf et al.

(1981), it can be shown that markedly different allocations arise under the linear logistic model.

$$\log \frac{p_{ij}}{1 - p_{ij}} = \mu + \alpha_i + \beta_j + \delta_{ij}. \tag{5-2}$$

Consequently, in the absence of reliable data on which model to choose, it would be advisable to search for a robust design that, although not fully optimal for any one model, would perform reasonably well under various plausible models.

Response-Surface Analysis

The use of a minimal four-point design with binary mixtures will provide little information on how the rate of occurrence of adverse health effects varies with the magnitude of exposure to both C_1 and C_2. The situation can be remedied with a more extensive two-factor factorial design in which k_1 and k_2 are greater than 1. The problem with this approach lies in the resource requirements for implementing a complete factorial experiment. With $k_1 = k_2 = k$, $(k + 1)^2$ treatment combinations will be required. Thus, even with only $k = 3$ doses of each component of the mixture, 16 combinations are needed if all possible combinations of the two components are used.

Another approach to investigating dose-response relations is to view the rate of occurrence of toxic responses to joint exposure to two materials as a surface with peaks (high response rates) and valleys or troughs (low response rates). Within this framework, the search procedures already designed for exploring response surfaces might be useful in elucidating responses to mixtures. Designs now exist for finding maximums (or minimums) in a response surface. However, the designs involve sequential administration of many dose combinations and the assumption that the response to a new combination can be determined quickly. Such designs are impractical for long-term toxicity studies or for experiments in which the use of a moderate number of treatment groups is economically infeasible. For short-term studies, however, such designs merit consideration. These include the acute-mortality studies with rats exposed to CO and CO_2 described in Appendix G.

Experimental designs for investigating response surfaces have been investigated in detail for use in nontoxicologic applications (Cornell, 1981). Much of this work focuses on the case of a continuous, rather than discrete, response and assumes that the response surface can be represented by a low-order polynomial, such as linear or quadratic. It is also generally assumed that the response of interest depends only on the relative proportions of the mixture components and not on the absolute amounts. Because the latter restriction is not acceptable in toxicologic research, the recent work by Piepel and Cornell (1985), which avoids this assumption, is of interest.

Much of the standard work on response-surface designs originated in industry. The responses of interest are usually clearly specified, a few properties of the mixture are considered, and the mixture ingredients are in some sense uniform. Examples include experiments with gasoline-blending and mixtures of food ingredients. One could argue that flour, sugar, and shortening are not chemically similar and that baking a cake does not involve interactions within a complex biologic system. A more relevant example is the blending of three pesticides to produce maximal toxicity in an insect species (Cornell, 1981); this example is interesting, because the physical, rather than chemical, properties of the constituents (one powder and two liquids) were important in mapping the response surface.

Two designs traditionally used in surface analysis are the simplex-lattice and simplex-centroid designs (Cornell, 1981). Only the relative proportions of the ingredients are assumed to affect the response, so the design space for a mixture of q components is a simplex of $q - 1$ dimensions. If x_i denotes the proportion of the ith component included in a particular treatment combination ($i = 1, \ldots, q$), the simplex consisting of all points $(x_1, \ldots, x_q)$ with $0 \leq x_i \leq 1$ and $x_1 + \ldots + x_q = 1$ represents all possible treatment combinations available for use in the experimental design. The simplex-lattice designs place all points on the boundary of the simplex, where at least one value of x_i is zero. In addition to such boundary points, the simplex-centroid designs include a point inside the boundary at the center of the simplex where all components are present in equal proportions (i.e., $x_i = 1/q$ for $i = 1, \ldots, q$). These designs are appropriate for minimizing the variances of the estimated regression coefficients in a polynomial response-surface model.

Response-surface designs have also been developed for studying specific regions within the design simplex. For example, a neighborhood around a particular mixture of current interest might be explored with central composite or rotatable designs (Carter et al., 1983). Alternatively, there might be upper or lower limits on the relative concentrations of some components. Multiple constraints on mixture components can lead to complex polygonal subregions inside the simplex, and design points are typically placed on the vertices, edges, and centroids of higher-dimensional faces of such subregions.

In considering the use of response-surface analysis in toxicologic applications, one must resolve several technical problems noted previously. In addition, the adequacy of statistical models for complex mixtures in toxicology is questionable. In this context, the question of interaction or synergism appears to be much more complex; in fact, there seems to be no general agreement as to models for representing the various alternatives, or even as to terminology (Kodell, 1986).

As a first approach, the natural additive model that is linear in the components may be considered for toxicologic applications. This model is both dose-additive and response-additive. Furthermore, the relative potency of two constituents is simply expressed as the ratio of the corresponding regression coefficients. Departures from additivity would indicate interaction among the components and may be represented by quadratic or cubic polynomial models. However, the assumption of a linear response function might not be reasonable in many situations.

Nonlinear models for continuous responses have been proposed by Ashford (1981) and Ashford and Cobby (1974), whereas Hewlett and Plackett (1959), Plackett and Hewlett (1967), and Christensen and Chen (1985) have discussed the quantal-response case. The optimal placement of design points seems to depend on the degree of nonlinearity of the response surface. Finney (1971) addressed the design question in the case of probit analysis and concluded that the optimal design points are intermediate between the median and extremes of the tolerance distribution. That conclusion suggests that simplex designs that put most of the design points on the boundary might not be appropriate for quantal-response models. Further investigation is needed.

Designs for Predicting Low-Dose Risks

Optimal designs for low-dose extrapolation with single chemicals require more information at low doses, because it is desirable to have more information at or near the doses about which we wish to make inferences and because the biologic responses at low doses might be different from those at high doses. This implies that more doses are needed at the low end of the dose-response curve, and greater weight will be placed on these doses, with respect to the number of animals to be assigned there.

Optimal designs for low-dose extrapolation with studies involving a single substance have been the subject of several investigations (Portier and Hoel, 1983; Krewski et al., 1984, 1986). If $P(d)$ denotes the probability of a toxic response occurring at dose d, the excess risk over background can be given by $\Pi(d) = P(d) - P(0)$. The experimental designs developed in these investigations are constructed to minimize the variance of the estimator of the excess risk at some fixed dose within the low-dose region based on a particular dose-response model.

Optimal designs for single chemicals have been developed for the probit, logit, Weibull, gamma multihit, and multistage models. These designs generally involve as many doses as there are parameters in the model, including a control group receiving zero exposure and a high-dose group capable of inducing a relatively high response rate. The intermediate doses tend to be assigned more animals than the control and high-dose groups.

Optimal designs for binary mixtures could be similarly developed to minimize the variance of the estimated excess risk $\Pi(d_1, d_2)$ at doses d_1 and d_2 of chemicals C_1 and C_2. Although these designs remain to be developed, we are able to provide some preliminary results here for particular special cases. Specifically, we consider the four-parameter logistic model in Equation E-36 of Appendix E. This model is used not because of any particular biologic appeal, but rather because it involves only four parameters (the minimum required to represent the background response rate, the effects of the two chemicals given separately, and their interaction). Rather than extrapolate directly from high to low doses, we use the model-based linear extrapolation procedure proposed by Van Ryzin (1980) for the case of single chemicals, as described in Appendix E. With binary mixtures, this involves extrapolation along a straight line joining the point $\Pi(d_1^*, d_2^*) = \pi^*$ and the origin, where (d_1^*, d_2^*) represents any combination of doses leading to an excess risk of π^* based on the fitted four-parameter logistic model.

Because only four parameters are involved, the optimal design will involve only four treatment combinations. For purposes of illustration, we take these to be given by $(d_1, d_2) = (0,0)$ (1,0), (0,1), and (1,1) as the 2 by 2 factorial designs discussed previously. This being the case, it remains only to determine the optimal allocation to these four groups.

The optimal allocations to these four cells with $\pi^* = 0.01$ are shown in Table 5-3 for selected values of tumor-response probabilities p_{ij}, with C_1 and C_2 assumed to act independently in the logistic model. (Here, we have selected $(d_1^* = d_2^*)$ to simplify the presentation, although this need not be the case.) These results indicate that, as the spontaneous-response rate p_{00} decreases, the fraction of subjects c_{00} assigned to the control group increases. The most interesting property of these designs is the relatively low-weight c_{11} assigned to the (1,1) cell in which C_1 and C_2 are administered jointly. The fact that this design places most weight on the cells in which C_1 and C_2 are administered separately might simply reflect the near additivity of the excess risks at the point (d_1^*, d_2^*).

To explore this point further, optimal designs for extrapolating from an ex-

TABLE 5-3 Optimal Experimental Designs Based on a Logistic Model for Low-Dose Extrapolation

Response Probabilities			Doses	Optimal Allocation[b]		
p_{00}	$p_{01} = p_{10}$	p_{11}[a]	$d_1^* = d_2^*$	c_{00}	$c_{01} = c_{10}$	c_{11}
			$\pi^* = 0.01$			
0.01	0.2	0.86	0.105	0.84	0.07	0.03
0.05	0.2	0.54	0.059	0.41	0.28	0.03
0.10	0.2	0.36	0.060	0.29	0.34	0.03
			$\pi^* = 0.10$			
0.01	0.2	0.86	0.340	0.60	0.10	0.21
0.05	0.2	0.54	0.305	0.25	0.27	0.21
0.10	0.2	0.36	0.322	0.19	0.31	0.19

[a] $p_{11} = \phi/(1 + \phi)$ where $\phi = (p_{10}p_{01}q_{00})/(q_{10}q_{01}p_{00})$.
[b] $c_{ij} = n_{ij}/n$.

cess risk of $\pi^* = 0.10$ were also calculated (Table 5-3). In this case, greater weight is assigned to the (1,1) cell, because additivity of the excess risks for C_1 and C_2 no longer holds at this point.

DESIGNS FOR INITIATION-PROMOTION STUDIES

Considerable evidence suggests that chemical carcinogenesis sometimes involves an initial irreversible change in target cells (initiation) that is followed by a partially reversible developmental phase (promotion), which leads to tumor expression. Initiation is thought to involve direct interaction between the proximate carcinogen and cellular DNA, whereas promotion is considered to be nongenotoxic. Early evidence of this mechanism was provided by Berenblum and Shubik (1947a,b, 1949), who applied a single noncarcinogenic dose of a polycyclic hydrocarbon to a mouse's skin and then administered croton oil repeatedly until tumors appeared. Similar results have since been obtained in other tissues, including rodent liver (Pitot and Sirica, 1980).

A biologically motivated mathematical model of carcinogenesis that incorporates the concepts of initiation, promotion, and progression is provided by the two-stage birth-death-mutation model developed by Moolgavkar, Venzon, and Knudson (Moolgavkar and Venzon, 1979; Moolgavkar and Knudson, 1981). The model assumes three possible fates for normal stem cells: death, division into normal progeny, and mutation resulting in one normal daughter cell and an intermediate or initiated cell. The population of initiated cells can similarly either divide, die, or undergo a second mutation to produce a fully transformed malignant tumor cell. In the context of this model, an initiator is a substance that increases the rate at which the first mutation occurs, whereas a promoter increases the pool of initiated cells available for malignant transformation. The term "progressor" has been used to describe a substance that increases the rate of the second mutation associated with the transformation of an initiated cell to a cancerous cell.

In examining a particular initiator-promoter pair, several treatment combinations might be required (Gart et al., 1986). To confirm a postulated initiation-promotion system, it might be necessary to include additional groups subjected to initiation but not promotion and to promotion but not initiation. Lifetime administration of initiator and promoter can also be considered to exclude the possibility that either one alone is acting as a complete carcinogen.

Other experimental protocols have involved the application of a second initiator after the administration of the original initiator-promoter pair, to study effects on the rate of occurrence of the second mutation (Scherer et al., 1984). This so-called IPI protocol can result in a marked increase in the production of neoplastic lesions, compared with the more conventional IP protocol, in which the second initiator is absent. In terms of experimental design, a combination of IP and IPI exposure regimens may be required for differentiation of effects on the two mutation rates involved in the two-stage birth-death-mutation model.

Practical Implications for Experimental Design

If the identification of a carcinogen in a long-term animal experiment is followed by another long-term animal experiment to develop a firm dose-response curve, at least 5 or 6 years will elapse from the start of the study to the development of the dose-response curve. Because such a long time might imply inadequate protection of the public health, this sequence has rarely (if ever) been followed. In its place, attempts have been made to use the data developed in the screening experiments as a basis for developing a dose-response relationship. It appears reasonable to develop omnibus experimental designs that can be used for both screening and predicting low-dose dose-response relationships.

Such hybrid designs will necessarily involve departures from optimal designs for carcinogen identification or dose-response curve development. In an attempt to identify a nearly optimal design that will perform reasonably well for purposes of both screening and low-dose extrapolation, the committee evaluated the efficiency of different four-point designs. Because of the low weight assigned to (1,1) cells in the optimal extrapolation designs, these designs are relatively inefficient for screening studies. Conversely, the screening designs performed better when used for extrapolation. A balanced design (i.e., equal numbers of animals per treatment group) appears to provide fairly good efficiency for both screening and extrapolation and might provide a reasonable solution to the problem of finding a suitable four-point hybrid design.

Although useful as a benchmark for evaluating other designs, the use of four-point optimal designs in experiments with two substances can be criticized on several counts. First, with only four treatment groups, no degrees of freedom are left over to assess the goodness of fit of a four-parameter model, such as the logistic model. Second, the loss of even a single dose group—for example, because of intercurrent mortality or because the MTD is exceeded—will result in a design with too few points for useful analysis. Third, additional treatment combinations involving joint exposure to various doses of the two substances of interest will be required, to yield some idea of the shape of the underlying response surface without relying on a particular parametric model.

Additional doses can be introduced into the design in a systematic way. For example, two materials can be given in all combinations of some maximal dose (D) of each material, half the maximal dose ($D/2$), one-fourth the maximal dose ($D/4$), and controls (0), yielding a total of 16 treatment combinations for testing two materials jointly. This rectangular design for testing two materials jointly will consume the resources necessary to test four materials one at a time, although the number of animals at a dose can be reduced or fractional factorial designs could be considered.

Because toxicity problems are likely to develop at combinations of high doses, the rectangular designs might become wasteful and fail to yield any usable response information at combinations of doses d_1 and d_2 in which both d_1 and d_2 are high. Exploration of other designs based on attempts to develop equitoxicity for combinations of exposures are thus certainly in order.

Other possibilities to be considered include test schemes for combinations in which the dose of one substance is fixed and the dose of another substance varies. This is an appealing procedure for materials to which the range of human exposure is rather narrow, so that exposures outside this range are rare enough to be uninteresting in a practical sense.

When the number of chemicals in combination is over two, fractional factorial designs should be considered. Such designs are thought useful when the effects of several factors are to be studied jointly. They permit estimation of the individual effects of each factor and provide information about interaction among factors, on the basis of results of tests with relatively few distinct laboratory mixtures or blends. The efficiencies of such designs hinge on the absence of high-order interactions, but first screenings that look for only low-order interactions and main effects might be well served by a design that places only relatively few animals at each of a specified number of points.

FUTURE DIRECTIONS

This chapter provides some ideas for assessing the toxicity of complex mixtures, but many of these ideas need further development and testing, and considerable progress needs to be made before definitive approaches can be recommended. This section highlights some of the research needs.

Cancer Models

The models developed to predict cancer risk associated with exposure to multiple agents are based on a general theory of the biologic mechanisms of carcinogenesis. These models predict that, even if extensive synergism takes place at exposures that lead to a directly observable carcinogenic response, one could still expect that at much lower exposures, response would not be distinguishable from additivity. This conclusion is based here primarily on a theoretical argument (see Chapter 2 for related human data). If the idea of low-dose additivity is to be given general credence, it is necessary to obtain evidence demonstrating that newly derived mathematical models can predict the joint effect of agents.

Some evidence that is amenable to new experimentation could be generated relatively quickly and inexpensively. For example, joint-exposure experimentation might be focused on for mutagenicity, cell proliferation, and DNA-adduct formation. The general approach would be to predict the joint response to agents on the basis of results obtained with single agents. The predictions would be verified by comparison with observed experimental results.

The key in applying a model to a particular chemical mixture is to postulate a specific mechanism of action for each component of the mixture and to use whatever information is available on the components to estimate values of the model parameters. Validation of this approach will require application of the model to available data sets and comparison of the risk predictions obtained with parameter estimates from short-term bioassays or other indirect measures of response with the results of chronic bioassays and ultimately epidemiologic results. For most mixtures, much of the information necessary for truly critical analysis of the postulated mechanisms of carcinogenesis is missing. However, validation of a model with available information is crucial and will be useful for directing future experimentation with mixtures.

Currently postulated mechanisms of carcinogenesis include the effects of chemicals on several biologic phenomena, such as DNA replication, cell proliferation, cellular toxicity, DNA-adduct formation, and mutagenicity. Although it has been difficult to incorporate this knowledge into risk assessments for single materials, it is worth attempting to examine the effects of chemical mixtures of unknown carcinogenicity on these phenomena. Establishing dose-response curves can provide some insight into whether a mixture is carcinogenic, the mechanism of such carcinogenicity, and how the mixture can be expected to act at low doses.

To predict whether interactions will occur and how a mixture might behave in environmental exposures, short-term bioassays—such as cell transformation, mutation, and proliferation—might help to determine the potential mechanisms of action of the mixture or its constituents and the shape of the dose-response curve (see Chapter 3). In the absence of chronic carcinogenesis bioassay data, mathematical models can be derived to describe the dose-response relationships observed in short-term bioassays and used to predict the carcinogenesis dose-response relationship (Thorslund et al., 1987). Research should be directed to the use of short-term bioassays to establish possible relationships.

Developmental Effects and Other Noncancer End Points

The research approach here is similar to that proposed by Murphy (1978): Postulate underlying biologic mechanisms, infer statistical models, test (to the extent possible) the reasonableness of the models with experimental and perhaps even epidemiologic data, and repeat the process as necessary and if possible. Until there is greater understanding of the biologic mechanisms underlying these end points, the development of suitable models will be hampered. In the interim, perhaps several candidate models could be examined for consistency with available data.

Developmental-effects experiments require fewer resources than cancer bioassays, and perhaps more thought should be given to using these experiments to identify potential interactive effects. The design of such experiments, however, needs development particularly with respect to exposure to mixtures and their components.

Statistical Approaches for Situations with No Preferred Dose-Response Model

In Appendix D several models have been applied to develop the toxicity estimates of a simple mixture. Research efforts include defining the set of appropriate models to apply in a bounding exercise. Several additional data sets could be examined in conjunction with this approach to help to determine its reasonableness and to identify a subset of models that appear to perform best. This approach might have implications for experimental design; these need to be investigated.

Empirical Modeling of the Toxicity of Mixtures

Generalized linear models are largely new to biologists and toxicologists, and there is a need to acquaint them with these models. More examples of their applicability should be sought with existing data sources.

Experimental design and generation of data for generalized linear models need more attention. Some candidate methods were discussed earlier; they need to be explored further.

As mixtures become more complex, potential independent variables in generalized linear models become more numerous. Research is needed to help to select the appropriate variables for these models.

Models that are based to a limited extent on biologic theory, such as the Ashford-Cobby model, might be useful, but they are awkward to apply (see Appendix G). Software is needed to facilitate the use of these models. The software and the models need to be extended beyond two-substance mixtures.

Pharmacokinetics

The role of pharmacokinetics in assessing the toxicity of single agents is increasingly recognized, but pharmacologic models that allow for interactions between agents are lacking and need to be developed. They could account, for example, for responses to mixtures of carcinogens with different mechanisms of action, such as initiators and promoters.

Methods also need to be developed to incorporate pharmacokinetic studies into chronic bioassays. Examination of drug studies might help in the design of such methods. Fractional factorial designs, often applied to chemical mixtures, should sometimes be incorporated into pharmacokinetic studies.

Pharmacokinetics might have a major role in testing the toxicity of mixtures, but the types of pharmacokinetic information that would be most useful for studying mixture toxicity are not yet well defined. Such information must be reconciled with the feasibility of estimating values of pharmacokinetic parameters.

Design of Experiments

This chapter has emphasized binary mixtures. More work is needed to extend some of the results with binary mixtures to more complex mixtures.

Because of toxicity problems that might arise with joint high-dose exposures, perhaps such combinations should be eliminated or de-emphasized in experimental designs involving complex mixtures. Ways to deal with this problem by formulating alternative designs could be usefully explored.

The current needs in response-surface exploration include techniques to reduce the number of dose combinations thought necessary to characterize a surface; ways to identify peaks and cliffs, which are expressions of nonadditive behavior among materials; and fractional factorial designs, which have been used with some success in agricultural research. Short-term stepwise or sequential testing techniques are clearly needed. Such techniques need to be interactive—that is, the results of experimentation should drive the next experimental steps and ultimately lead to the creation of new models. Some techniques have been applied to nonbiologic problems; their use with toxicology data should be explored.

Strategy Definition

Several strategies for testing complex mixtures are suggested elsewhere in this report. Formal mathematical modeling has only partially encompassed these proposed strategies, and additional development has been left for further research. These strategies include the following:

- A multistep (tier) approach and a matrix approach of the type indicated here for two-material comparisons, including a process involving critical-element variability, that is, varying one factor while another is held constant.
- Bioassay-driven fractionation of a complex mixture in which fractions of a mixture are evaluated in a standard bioassay to determine which components or combinations of components are responsible for the carcinogenicity (or other toxicity) of the mixture. Substantial work will be necessary to develop search and testing strategies that will minimize the effort required to identify the toxic fractions (and their constituents).

- Analysis of the complex mixture to identify (known) toxic components and testing of various components (in combination) in an attempt to reproduce the toxicity of the parent mixture. No effective modeling or combining strategies have been developed to minimize the work necessary to "rebuild" a complex toxic material.

The last process mentioned should be capable of distinguishing noninteraction of components (expected joint-action behavior) from interaction of components (both synergism and antagonism) and, in the case of interaction, provide leads to the discovery of the mechanisms of interaction.

REFERENCES

Andersen, M. E., H. J. Clewell III, M. L. Gargas, F. A. Smith, and R. H. Reitz. 1987. Physiologically based pharmacokinetics and the risk assessment process for methylene chloride. Toxicol. Appl. Pharmacol. 87:185–205.

Ashford, J. R. 1981. General models for the joint action of mixtures and drugs. Biometrics 37:457–474.

Ashford, J. R., and J. M. Cobby. 1974. A system of models for the action of drugs applied singly or jointly to biological organisms. Biometrics 30:11–31.

Berenblum, I., and P. Shubik. 1947a. A new, quantitative approach to the study of the stages of chemical carcinogenesis in the mouse's skin. Br. J. Cancer 1:383–391.

Berenblum, I., and P. Shubik. 1947b. The role of croton oil applications, associated with a single painting of a carcinogen, in tumour induction of the mouse's skin. Br. J. Cancer 1:379-382.

Berenblum, I., and P. Shubik. 1949. The persistence of latent tumour cells induced in the mouse's skin by a single application of 9:10-dimethyl-1:2-benzanthracene. Br. J. Cancer 3:384–386.

Bliss, C. I. 1939. The toxicity of poisons applied jointly. Ann. Appl. Biol. 26:585–615.

Carter, W. H., G. L. Wampler, and D. M. Stablein. 1983. Experimental design, pp. 108–129. In Regression Analysis of Survival Data in Cancer Chemotherapy. Marcel Dekker, New York.

Chen, C. W., and J. N. Blancato. 1987. Role of pharmacokinetic modeling in risk assessment: Perchloroethylene (PCE) as an example, pp. 367–388. In Pharmacokinetics and Risk Assessment. Vol. 8. Drinking Water and Health. National Academy Press, Washington, D.C.

Christensen, E. R., and C.-Y. Chen. 1985. A general noninteractive multiple toxicity model including probit, logit, and Weibull transformations. Biometrics 41:711–725.

Cornell, J. A. 1981. Experiments with Mixtures: Designs, Models, and the Analysis of Mixture Data. John Wiley & Sons, New York. (305 pp.)

Dumouchel, W. H., and J. E. Harris. 1983. Bayes methods for combining the results of cancer studies in humans and other species. J. Am. Stat. Assoc. 78:293–315.

Finney, D. J. 1952. Probit Analysis: A Statistical Treatment of the Sigmoid Response Curve. 2nd ed. Cambridge University Press, Cambridge.

Finney, D. J. 1971. Probit Analysis. 3rd ed. Cambridge University Press, New York. (333 pp.)

Gart, J. J., D. Krewski, P. N. Lee, R. E. Tarone, and J. Wahrendorf. 1986. Statistical Methods in Cancer Research, Vol. III: The Design and Analysis of Long-Term Animal Experiments. IARC Scientific Publications No. 79. International Agency for Research on Cancer, Lyon, France. (213 pp.)

Gibaldi, M., and D. Perrier. 1975. Pharmacokinetics. Marcel Dekker, New York.

Harris, J. E. 1983a. Diesel emissions and lung cancer. Risk Anal. 3(2):83–100.

Harris, J. E. 1983b. Diesel emissions and lung cancer revisited. Risk Anal. 3(2):139–146.

Hewlett, P. S., and R. L. Plackett. 1959. A unified theory for quantal responses to mixtures of drugs: Non-interactive action. Biometrics 15:591–610.

Hoel, D. G., N. L. Kaplan, and M. W. Anderson. 1983. Implication of nonlinear kinetics on risk estimation in carcinogenesis. Science 219:1032–1037.

John, P. W. 1971. Statistical Design and Analysis of Experiments. Macmillan, New York. (356 pp.)

Kodell, R. L. 1986. Modeling the Joint Action of Toxicants: Basic Concepts and Approaches. Presented at the ASA/EPA Conference on Interpretation of Environmental Data: The Current Assessment of Combined Toxicant Effects, Washington, D.C., May 5-6, 1986. (21 pp.)

Kodell, R. L., and J. G. Pounds. In press. Assessing the toxicity of mixtures of chemicals. In D. Krewski and C. Franklin (eds.). Statistical Methods in Toxicological Research. Gordon and Breach, New York.

Krewski, D., J. Kovar, and M. Bickis. 1984. Optimal experimental designs for low dose extrapolation. II. The case of nonzero background, pp. 167–191. In Y. P. Chaubey and T. D. Dwivedi (eds.). Topics in Applied Statistics. Proceedings of the Statistics '81 Canada Conference held at Concordia University, Montreal, Quebec, April 29–May 1, 1981.

Krewski, D., M. Bickis, J. Kovar, and D. L. Arnold. 1986. Optimal experimental designs for low dose extrapolation. I. The case of zero background. Utilitas Math. 29:245–262.

Laird, N. M., and T. A. Louis. 1986. Combining data from different sources: Empirical Bayes confidence intervals. Paper presented at the XIIIth International Biometric Conference, Seattle, Washington, July 29–August 1, 1986.

Moolgavkar, S. H., and A. G. Knudson, Jr. 1981. Mutation and cancer: A model for human carcinogenesis. J. Natl. Cancer Inst. 66:1037–1052.

Moolgavkar, S. H., and D. J. Venzon. 1979. Two-event models for carcinogenesis: Incidence curves for childhood and adult tumors. Math. Biosci. 47:55–77.

Murphy, E. A. 1978. Epidemiological strategies and genetic factors. Int. J. Epidemiol. 7:7–14.

National Toxicology Program, Board of Scientific Counselors. 1984. Report of the NTP Ad Hoc Panel on Chemical Carcinogenesis Testing and Evaluation. U.S. Government Printing Office, Washington, D.C. (280 pp.)

Piepel, G. F., and J. A. Cornell. 1985. Models for mixture experiments when the response depends on the total amount. Technometrics 27:219–227.

Pitot, H. C., and A. E. Sirica. 1980. The stages of initiation and promotion in hepatocarcinogenesis. Biochim. Biophys. Acta 605:191–215.

Plackett, R. L., and P. S. Hewlett. 1948. Statistical aspects of the independent joint action of poisons, particularly insecticides. I. The toxicity of a mixture of poisons. Ann. Appl. Biol. 35:347–358.

Plackett, R. L., and P. S. Hewlett. 1967. A comparison of two approaches to the construction of models for quantal responses to mixtures of drugs. Biometrics 23:27–44.

Portier, C., and D. Hoel. 1983. Optimal design of the chronic animal bioassay. J. Toxicol. Environ. Health 12:1–19.

Ramsey, J. C., and M. E. Anderson. 1984. A physiologically based description of the inhalation pharmacokinetics of styrene in rats and humans. Toxicol. Appl. Pharmacol. 73:159–175.

Scherer, E., A. W. Feringa, and P. Emmelot. 1984. Initiation-promotion-initiation. Induction of neoplastic foci within islands of precancerous liver cells in the rat. In M. Borśonyi, K. Lapis, N. E. Day, and H. Yamasaki (eds.). Models, Mechanisms and Etiology of Tumour Promotion. IARC Scientific Publications No. 56. International Agency for Research on Cancer, Lyon, France.

Thorslund, T. W., C. C. Brown, and G. Charnley. 1987. Biologically motivated cancer risk models. Risk Anal. 7:109–119.

Van Ryzin, J. 1980. Quantitative risk assessment. J. Occup Med. 22:321–326.

Verducci, J. S. 1985. Task 61: Chains of Extrapolation. EPA Contract No. 68-01-6721. Report to EPA Office of Pesticides and Toxic Substances. Batelle Columbus Division, Washington Operations, Washington, D.C. (38 pp.)

Wahrendorf, J., R. Zentgraf, and C. C. Brown. 1981. Optimal designs for the analysis of interactive effects of two carcinogens or other toxicants. Biometrics 37:45–54.

Whittemore, A. S., S. C. Grosser, and A. Silvers. 1986. Pharmacokinetics in low dose extrapolation using animal cancer data. Fundam. Appl. Toxicol. 7:183–190.

APPENDIXES

A

Origins of Complex Mixtures

COMBUSTION AND DISTILLATION PRODUCTS

Combustion and distillation products can consist of thousands of chemicals from fossil- and synthetic-fuel sources, vegetable sources, or synthetic materials. These mixtures generally contain a high percentage of organic chemicals, including polycyclic aromatic compounds and aliphatic compounds, inorganic materials, and metallic and nonmetallic salts. The mixtures can be heterogeneous and consist of both volatile gases (e.g., CO, NO_x, SO_2, and formaldehyde) and semivolatile materials (e.g., two- and three-ring aromatic compounds) and condensed organic and inorganic matter adsorbed on carbonaceous material.

Fossil Fuels

Coal

The opportunity for synthesis of hazardous chemicals exists whenever coal is subjected to severe conditions, such as those of pyrolysis in the production of coal tar and coal-tar pitch and those of hydrogeneration or gasification in the production of synthetic fuels (McNeil, 1966).

In general, the physical characteristics of coal-tar pitch are measurable and predictable. In chemical composition, coal-tar pitch is a highly complex organic material that varies from batch to batch, depending on temperature and refining procedures. Polycyclic aromatic hydrocarbons (PAHs) are important components of coal tar and coal-tar pitch and have been used as predictors of potential hazards (Gray, 1984; Richards et al., 1979).

Petroleum

Of the various petroleum fractions, asphalt, gasoline, heavy distillates, and diesel fuels appear to be of most interest to the toxicologist (MacFarland et al., 1984). The biologic activity of petroleum fractions does not correlate with aromatic content, but rather might be due to the promoter activity of the aliphatic fraction. Extracts of diesel-exhaust particles have been found to contain aliphatic hydrocarbons, PAHs, nitroaromatics, and oxygenated PAHs; much of the toxicologic research has focused on the nitroaromatics. The heavy distillates and residues from catalytic cracking have been found to contain PAHs and saturated hydrocarbons, which contribute to the carcinogenic activity of the materials.

Oil Shale

Oil shale undergoes thermal decomposition processes similar to those of coal. The possibility of the synthesis of hazardous chemicals during pyrolysis procedures exists. Raw shale and spent shale are not biologically active, but the shale oil produced has been found to be very carcinogenic (Barkley et al., 1979a,b; Bingham and Barkley, 1979). Shale-oil retort samples have been found to contain a variety of PAHs in parts-per-billion concentrations. The biologic activities of shale oils do not correlate with the concentrations of benzo[a]pyrene in the samples.

Synthetic Fuels

Major processes that have been used for coal liquefaction are pyrolysis, hydrogenation, solvent refining, and Fisher-Tropsch synthesis. All those processes involve harsh conditions that produce hazardous chemicals. All the products of hydrogenation—high-boiling-point distillates, centrifuged oils, char, residues, recycled solvent oil, recycled solvent, and liquid coal—are potentially hazardous. The synthetic coal liquids from all processes are higher in PAH, basic, acidic, and insoluble fractions than are crude petroleum products. The higher-boiling-point cuts appear to be enriched in PAHs and nitrogen- and oxygen-containing molecules (Beland et al., 1985; Gray, 1984; Pelroy and Wilson, 1981).

However, synthetic gas is not expected to pose a carcinogenic risk. Trace elements, organic carcinogens, or cocarcinogens present can be removed from the raw product during cleanup and scrubbing. In coal gasification, the conversion process itself, rather than the fuel produced, should be the primary concern. Potentially carcinogenic polycyclic organic material found during gasification of coal is likely to concentrate in the tars, oils, char, quench waters, and organic substances from gas-entrained particles. In addition, if the crude gas contains tars of high-boiling-point oils, they must be considered hazardous. The tars contain high-molecular-weight polyaromatic species, which appear to pose the greatest health concern (Richards et al., 1979).

Vegetable Sources

Tobacco and wood are two major vegetable combustion sources. Tobacco smoke is one of the most extensively studied complex mixtures. Hydrogenation, pyrolysis, oxidation, decarboxylation, and dehydration are all involved in tobacco combustion. The vapor phase of cigarette smoke contains nickel carbonyl, hydrazine, vinyl chloride, formaldehyde, and inorganic gases, such as CO_x, NO_x, and HCN (Hoffmann and Wynder, 1976; Hoffmann et al., 1976). Cigarette-smoke condensate (CSC) contains aliphatic hydrocarbons, aromatic hydrocarbons, phenols, and long-chain acids and alcohols, some of which are cocarcinogenic (Van Duuren and Goldschmidt, 1976), tumor-promoting, and tumor-inhibiting agents. Nitrosamines are present in sidestream smoke.

Thus, cigarette smoke contains many biologically active compounds. A variety of fractionation procedures have been used to identify these components, but, despite elaborate analytic procedures, it has been difficult to design toxicologic strategies that address the many biologically active compounds present in this very complex mixture.

Pyrolysis products of vegetable materials also have attributes in common with materials isolated from cooked food.

Synthetic Materials

Combustion of synthetic materials—such as polyurethane foam, polyester, polyethylene, polystyrene, and polyvinyl chloride—results in the production of many components (Levin, 1986). Depending on the synthetic material used, thermal-degradation products include aliphatic and aromatic hydrocarbons, aliphatic amines, aldehydes, ketones, acids, and a mixture of toxic gases. The acute toxicity of a material's combustion products can be explained by either the interactions of major gases (e.g., CO, CO_2, and HCN) or the primary thermal-degradation products.

NONCOMBUSTED MATERIALS

Coal Dust

Coal can be described as a compact stratified mass of vegetation, interspersed with smaller amounts of inorganic matter, that has been modified chemically and physically by agents over a very long time. The chemical properties of coal depend on the amounts and ratios of the vegetation's constituents, the nature and quantity of inorganic material, and the changes that the constituents have undergone. Coal has a complicated chemical structure based on carbon and hydrogen with various amounts of oxygen, nitrogen, and sulfur. Bituminous coal, from which coal-tar pitch is derived, contains a number of PAHs and toxic trace elements (Francis, 1961; Torrey, 1978).

Oil Shale

Health and environmental concerns have been raised in connection with oil-shale extraction and processing. Blasting and mining produce dust, particulate organic matter (POM), gases, hydrocarbons, silica, and metal salts. Crushing and screening produce more dust, silica, and POM. Retort operations can also produce polycyclic organic compounds, H_2S, NH_3, and volatile substances. Some of those compounds, as well as arsenic and other metals, are produced in the upgrading process, and disposal of the solid waste products is a major environmental concern. Many of the compounds can contaminate both the atmosphere and water. Because water is used in the oil-shale industry, many of the compounds are found in the waste and runoff waters. Many potentially hazardous chemicals might be present in the work area and environs. The major health concern appears to be the carcinogenicity associated with fossil-fuel production (Barkley et al., 1979a,b).

Waste

Disposal of solid wastes produced in processing and extracting of fossil fuels presents a major environmental problem. For example, the potential hazard to workers who handle spent shale, which can contain substantial quantities of silica and PAHs, should be of primary concern. The possibility that carcinogens will leach into the environment is also a concern. Spent shale might be used in revegetation of disturbed land, so it is possible that products leached from spent shale will be ingested by plants containing the contaminants.

Dump sites can contain hazardous chemicals from various sources. These materials are usually very complex mixtures with nonpredictable chemical composition. The source of materials at the dump site is often unknown, so the analytic procedures are even more difficult. Municipal waste and sewage waste, although complex, are somewhat more predictable.

Water

Raw and finished surface and ground drinking waters contain carcinogenic, mutagenic, and toxic chemicals. Industrial and municipal discharges, urban and rural runoff, natural sources, and water and sewage chlorination practices have been identified as possible sources of those pollutants (Menzer and Nelson, 1986). In a surface drinking-water sample, approximately 460 organic compounds were identified, including 41 PAHs, 15 PCBs, and some amines, amides, and halogenated species (Coleman et al., 1980), as well as a number of inorganic chemicals. Knowledge and understanding of basic chemical principles is important in predicting the chemical composition of drinking water. For example, chlorine, chlorine dioxide, and chloramine treatment processes have been used for drinking-water disinfection. Those practices have generated a number of disinfectant byproducts from the reactions between background organic chemicals and chlorine.

REFERENCES

Barkley, W., D. Warshawsky, and M. Radike. 1979a. Toxicology and carcinogenicity of oil shale products, pp. 79–95. In O. White, Jr. (ed.). Proceedings of the Symposium on Assessing the Industrial Hygiene Monitoring Needs for Coal Conversion and Oil Shale Industries. Brookhaven National Lab, Upton, N.Y. (Available from NTIS as BNL-51002.)

Barkley, W., D. Warshawsky, R. R. Suskind, and E. Bingham. 1979b. Toxicology and carcinogenic investigation of shale oil and shale oil products, pp. 157–162. In Oak Ridge National Laboratory. Proceedings of the Symposium on Potential Health and Environmental Effects of Synthetic Fossil Fuel Technologies, Gatlinburg, Tenn., September 25, 1978. Oak Ridge National Laboratory, Oak Ridge, Tenn. (Available from NTIS as CONF-780903.)

Beland, F. A., R. H. Heflich, P. C. Howard, and P. P. Fu. 1985. The *in vitro* metabolic activation of nitro polycyclic aromatic hydrocarbons, pp. 371–396. In American Chemical Society Monograph 283. American Chemical Society, Washington, D.C.

Bingham, E., and W. Barkley. 1979. Bioassay of complex mixtures derived from fossil fuels. Environ. Health Perspect. 30:157–163.

Coleman, W. E., R. G. Melton, F. C. Kopfler. K. A. Barone, T. A. Aurand, and M. G. Jellison. 1980. Identification of organic compounds in mutagenic extract of a surface drinking water by a computerized gas chromatography/mass spectrometry system (GC/MS/COM). Environ. Sci. Technol. 14:576–588.

Francis, W. 1961. Coal: Its Formation and Composition. Edward Arnold, London. (47 pp.)

Gray, R. H. 1984. Chemical and toxicological aspects of coal liquefaction and other complex mixtures. Regulatory Toxicol. Pharmacol. 4:380–390.

Hoffmann, D., and E. L. Wynder. 1976. Environmental respiratory carcinogenesis, pp. 324–365. In C. E. Searle (ed.). Chemical Carcinogens. American Chemical Society Monograph No. 173. American Chemical Society, Washington, D.C.

Hoffmann, D., C. Patrianakos, K. D. Brunnemann, and G. B. Gori. 1976. Chromatographic determination of vinyl chloride in tobacco smoke. Anal. Chem. 48:47–50.

Levin, B. C. 1986. A Summary of the NBS Literature Reviews on the Chemical Nature and Toxicity of the Pyrolysis and Combustion Products from Seven Plastics: Acrylonitrile-Butadiene-Styrenes (ABS), Nylons, Polyesters, Polyethylenes, Polystyrenes, Poly(Vinyl Chlorides) and Rigid Polyurethane Foams. NBSIR 85-3267. National Bureau of Standards, Gaithersburg, Md.

MacFarland, H. N., C. E. Holdsworth, J. A. MacGregor, R. W. Call, and M. L. Lane. 1984. Applied Toxicology of Petroleum Hydrocarbons. Advances in Modern Environmental Toxicology, Vol. VI. Princeton Scientific Publishers, Princeton, N.J. (287 pp.)

McNeil, D. 1966. Coal Carbonization Products. Pergamon Press, New York. (159 pp.)

Menzer, R. E., and J. O. Nelson. 1986. Water and soil pollutants, pp. 825–853. In C. D. Klaasen, M. O. Amdur, and J. Doull (eds.). Casarett and Doull's Toxicology: The Basic Science of Poisons, 3rd ed. Macmillan, N.Y.

Pelroy, R. A., and B. W. Wilson. 1981. Fractional Distillation as a Strategy for Reducing the Genotoxic Potential of SRC-II Coal Liquids: A Status Report. PNL-3787. Pacific Northwest Lab, Richland, Wash. (Available from NTIS as DE 82002242.) (82 pp.)

Richards, D. E., W. P. Tolos, J. B. Lal, and C. V. Cooper. 1979. Tumors Induced in C3H/HeJ Mice by Coal Tar Neutral Subfractions. DHEW (NIOSH) Publication No. 80-101. National Institute for Occupational Safety and Health, Cincinnati, Ohio. (Available from NTIS as PB 80-175-854.)

Torrey, S., ed. 1978. Trace Contaminants from Coal. Pollution Technology Review No. 50. Noyes Data Corporation, Park Ridge, N.J.

Van Duuren, B. L., and B. M. Goldschmidt. 1976. Cocarcinogenic and tumor-promoting agents in tobacco carcinogenesis. J. Natl. Cancer Inst. 56:1237-1242.

B

Case Studies Establishing Active Agents and/or Interactions in Complex Mixtures

SULFUR DIOXIDE AND SUSPENDED PARTICULATE MATTER

The case of sulfur dioxide (SO_2) and particulate matter (PM) in ambient air and their effects on mortality and respiratory disease morbidity is a classic case of exposure to complex mixtures in which identification of causal factors has proved extremely difficult. Epidemiologic research has been critical in determining the relative roles of SO_2 and PM, whereas laboratory inhalation studies have been critical in identifying active components in the PM. An important caveat is that the critical experimental inhalation studies were done on humans and animals not normally used in standardized bioassays (i.e., dogs, rabbits, and guinea pigs). Also, some of the effects assays—particle clearance rates, detailed airway morphometry, and CO-diffusing capacity—are not used in conventional rodent bioassay tests. Unconventional animal models and non-routine effects measures were needed because available animal models for the human health effects of interest, such as chronic bronchitis and exacerbations of asthma, have not been widely accepted as relevant to human disease.

Air pollution health effects have been associated with coal smoke in London in a qualitative sense at least since the thirteenth century, when Edward I banned the use of coal during sessions of Parliament. However, sustained efforts to control coal-smoke pollution did not begin until the London "killer fog" of December 1952 (Great Britain, 1954). Because the effluent from coal combustion contains both SO_2 and PM, scientists long debated whether SO_2 or PM was the causal agent or whether the effects were due to their combined actions, or synergism. In the absence of clear evidence in the 1950s and 1960s, various authorities responsible for air pollution control made different judgments. In the United Kingdom, authorities made efforts to reduce PM, especially the black-smoke component, which was cut by about a factor of 2 between 1952 and 1962. The reduction in SO_2 over the same period was substantially smaller. Later reviews of the changes in mortality and morbidity indexes during this period tended to implicate PM as a more important health stressor than SO_2.

Time-series analyses of daily mortality in New York City (Schimmel, 1978; Schimmel and Murawski, 1975, 1976; Buechley et al., 1973) indicated that mortality increased with increasing SO_2 concentrations for both the periods 1963–1966 and 1970–1972. Because the absolute SO_2 concentrations dropped by about a factor of 3 over this period, it is highly unlikely that SO_2 was the causal factor for the similar association with mortality in the different periods. Rather, the daily fluctuations in SO_2 appeared similar to daily fluctuations in other pollutants from the same or similar sources, and other pollutants in the pollutant mixture were probably more directly related to the observed effect. This leads to the conclusion that PM or PM components are better indexes of exposure.

The active agents in PM have not yet been clearly identified. Ozkaynak and Spengler (1985) performed regression analyses between daily mortality and four different components of PM: total suspended particulate matter (TSP), inhalable particulate matter (IP), fine particulate matter (FP), and SO_4^{2-}. TSP has an upper cut-size of $\simeq$ 40–60 μm aerodynamic diameter, depending on wind speed and direction. IP has an upper cut-size of 15 μm, which is relatively independent of wind speed and direction. FP has an upper cut-size of 2.5 μm. SO_4^{2-} is almost all within the FP fraction and varies from less than 10% to more than 50% of FP. All four showed positive regression with daily mortality, with correlation coefficients increasing in the order of TSP $<$ IP $<$ FP $<$ SO_4^{2-}. Only FP and SO_4^{2-} had statistically significant coefficients.

Although SO_4^{2-} might be the best predictor of health effects among the commonly measured indexes of PM, it is probably not the active agent. Several physiologic effects of sulfate aerosols have been shown to depend on their H^+ content. These include specific airway conductance (Utell, 1985) and bronchial mucociliary particle clearance (Schlesinger, 1985). Thus, SO_4^{2-} might be acting as a surrogate for the strong acid associated with it. Unfortunately, correlation studies between H^+ and health effects cannot be performed, because there are too few data available on ambient H^+ concentrations (Lippmann, 1985).

Experimental inhalation studies with SO_2 and PM have not played any major role in elucidating the role of these pollutants in population mortality and morbidity experience. The only documented human health effect of SO_2 exposures at concentrations near those occurring in ambient air are transient changes in airway resistance and compliance. Amdur's guinea pig model (Amdur and Mead, 1958) is much more responsive to SO_2 than is any other animal tested

(U.S. EPA, 1982). In this respect, it is similar in response to asthmatic humans, who, in controlled chamber exposure studies, are about 10 times more responsive (Sheppard et al., 1980) than healthy humans. However, the transient bronchoconstrictive responses seen in these human and guinea pig studies are qualitatively different from those reported for human populations exposed to polluted ambient air.

Amdur's guinea pig model also has been used to demonstrate synergism between SO_2 and PM. The bronchoconstrictive effects of SO_2 are potentiated by coexposure to hygroscopic and catalytic aerosols (McJilton et al., 1976; Amdur and Underhill, 1968). However, both the end point used—transient bronchoconstriction—and the very high (> 5 mg/m^3) concentrations of particles needed to cause the potentiated response make the relevance of these responses to the effects associated with ambient pollutant exposures somewhat questionable.

Experimental PM inhalation studies with concentrations and compositions similar to those encountered in ambient air have generally been unproductive (U.S. EPA, 1982). The only constructive interpretation that one can make of the data from such studies is that insoluble materials—such as fixed carbon, resuspended soil, and fly ash—could not by themselves account for the effects observed.

The only experimental inhalation studies producing effects of possible relevance to the human experience have been those involving acidic aerosols, either alone or in complex mixtures. Some recent animal inhalation studies by Amdur (1985) demonstrated that effects produced by single exposures at very low acid concentrations can be persistent. She exposed guinea pigs by inhalation for 3 hours to the diluted effluent from a furnace that simulates a model coal combuster. The amount of H_2SO_4 on the surface of the ZnO particles was less than 40 μg/m^3. These aerosol studies produced significant decrements of total lung capacity (TLC), vital capacity (VC), functional residual capacity (FRC), and CO-diffusing capacity (DL_{CO}). At 12 hours after exposure, there were distension of the perivascular and peribronchial connective tissues and an increase in lung weight. The alveolar interstitium also appeared distended. At 1 hour, there was an increase in lung permeability. At 72 hours, TLC, VC, and FRC had returned to baseline values, but DL_{CO} was still significantly depressed. Based on her experience with pure SO_2 and pure H_2SO_4 exposures in the guinea pig model, Amdur concluded that the furnace effluent produced an acid aerosol effect because of its persistence. The persistent changes in function and morphological changes after exposure to acidic aerosol at very low concentrations suggested that repetitive exposures could lead to chronic lung disease.

Another animal inhalation study with implications for human chronic lung disease involved a group of beagle dogs exposed to 16 hours daily for 68 months to a mixture of SO_2 at 1,100 μg/m^3 and H_2SO_4 at 90 μg/m^3, either with or without diluted auto exhaust. The dogs were maintained without further exposure for an additional 3 years, during which there was a progression of effects of interest to the pathogenesis of chronic lung disease (U.S. EPA, 1980). Clinical examinations were conducted regularly throughout the 8.7-year experimental period. Small abnormalities in pulmonary function were detected after 61 months of exposure, 7 months before it ended. When the same dogs were studied 2 years after the termination of exposure, many of their pulmonary function values showed even greater differences from those of controls. Whereas the controls appeared to have stable pulmonary function values similar to those of other healthy beagles, morphologic and morphometric analysis revealed that the exposed dogs' lungs exhibited two important structural abnormalities: enlargement and breakdown of air spaces centered on respiratory bronchioles and alveolar ducts. The abnormality in air spaces can be considered analogous to early human centrilobular (centriacinar) emphysema. Although mild compared with clinically significant emphysema in humans, it was sufficient to cause abnormalities in pulmonary function. The other major structural abnormality was hyperplasia of bronchiolar epithelial cells. The structural abnormalities persisted for 3 years after the 68-month exposure. The indication from the continued loss of pulmonary function after exposure ceased is that the damage worsened. The conclusion that the long-term low-concentration exposures produced persistent damage that, although slight, appeared to be progressive makes the findings important in predicting possible effects in humans over a life span.

A further conclusion is that, because of the complexity of biological interactions in the lungs of an exposed person or animal, judiciously selected long-term studies will be necessary to fill the gap between short-term in vitro and in vivo experimental studies and epidemiologic observations of human populations.

Although single brief exposures to acidic aerosols can either accelerate or retard mucociliary particle clearance, depending on the dose distribution along the airways, the effects, which are similar in humans and animals, are transient. In these tests, H_2SO_4 was at least twice as effective per mole as NH_4HSO_4, whereas the more neutral salts $(NH_4)_2SO_4$ and Na_2SO_4 had no detectable effects on mucociliary clearance (Schlesinger, 1985). In followup tests involving repetitive daily exposures to H_2SO_4 at a concentration that, on single exposure, produced minimal transient effects on tracheobronchial mucociliary clearance function, there were highly variable clearance rates and persistent shifts from baseline clearance rates. After 20 days of exposures, rabbits exhibited an increased density of secretory cells and thickened epithelial layers in medium and small conducting airways. During repetitive daily exposures, there was an acceleration of early alveolar clearance during the first 2 weeks of exposure and a similar acceleration at 8–10 weeks of exposure. The progression, or possibly the regression, of these changes in clearance rate during further periods of daily exposures needs to be determined. We can expect these issues to be clarified as the results from longer exposure periods become available.

The significance of the changes in particle clearance rate in the lungs from repetitive daily exposures to acidic aerosols, in terms of the pathogenesis of chronic respiratory disease, is not yet clear. But the close correspondence between the effects of cigarette smoke and H_2SO_4 aerosol on mucociliary clearance after both short-term and long-term exposures, the similarities between the epithelial changes after repetitive H_2SO_4 inhalation in rabbits and those seen in the lungs of young smokers in postmortem examinations, and the well-established role of smoking in the etiology of chronic bronchitis combine to suggest that chronic bronchitis could result from long-term repetitive exposures to H_2SO_4 (Lippmann et al., 1982). The fact that the incidence of chronic bronchitis among nonsmokers was higher in the United Kingdom when ambient acidic aerosol concentrations were high lends further plausibility to a causal relationship. Much more evidence is needed to demonstrate a clear causal relationship, however.

Conventional inhalation studies with rodents have not been useful in unraveling the effects associated with human exposures to the ambient complex mixtures of sulfur oxides and particles. Research studies with less commonly used experimental animals (e.g., guinea pigs, dogs, and rabbits) have been very useful, but not yet definitive. Perhaps the major lesson is that studies focused on mechanisms of the pathogenesis of chronic lung disease need to be designed with care, with constant attention to the suitability of the animal model and of the relationship to the real-world question.

In summary, the excess mortality and morbidity that have been demonstrated in populations exposed to atmospheres containing mixtures of fossil-fuel combustion products might well have been due to a specific component of the mixture—for example, the strong acid component of the aerosol—or to the combined or synergistic effects of two or more of the many components in the mixture. A more definitive summary is not yet possible, although current laboratory and epidemiologic research might permit one in the near future.

LEAD AND NUTRITIONAL FACTORS—EFFECTS ON BLOOD PRESSURE

Hypertension has long been recognized as a risk factor for cardiovascular disease, and several environmental and nutritional factors have been shown to affect blood pressure in experimental and epidemiologic studies (U.S. EPA, 1984). Among environmental factors that have been associated with blood pressures are lead (Pb) and noise. Among dietary factors associated with blood pressure are calcium (Ca), zinc (Zn), phosphorus (P), alcohol consumption, and vitamins A and C. Although Pb is a single substance, it is always found in the environment as a part of a complex mixture. This case study illustrates how the important influence of blood Pb on blood pressure was recently established by epidemiologic studies. It also discusses why animal-study results, which could have been used to anticipate the effects seen in humans, were not given the attention or recognition they deserved.

The role of Pb as a pollutant stressor for increased blood pressure could well be confounded by the well-established role of Ca as a suppressor of blood pressure. It is possible that persons with high Ca consumption have both decreased blood pressure and reduced blood Pb due to the competition of both Pb and Ca for the same binding sites. The influence of the other cofactors known to affect blood pressure complicates the task of establishing the extent to which Pb constitutes an important risk factor for increased blood pressure.

A consistent pattern of results has begun to emerge from recent investigations of the relations between small Pb exposures and increases in blood pressure (hypertension). Khera et al. (1980) reported higher blood Pb concentrations in hypertensive patients and those with other cardiovascular diseases than in hospital control subjects. Kromhout and de Lezenne Coulander (1984) and Kromhout et al. (1985) reported associations between hypertension and blood Pb among elderly men in the Netherlands. Batuman et al. (1983) reported an association between hypertension and chelatable Pb burdens in veterans. Moreau et al. (1982) reported significant associations ($p < 0.001$) between blood Pb concentrations and a continuous measure of blood pressure among French police officers after controlling for important potential confounding variables, such as age, body-mass index, smoking, and drinking. And Weiss et al. (1986) reported that—when previous systolic blood pressure, body-mass index, age, and smoking were corrected for—a high blood Pb concentration was a significant predictor of later increase in systolic pressure in police officers in Boston.

In a recent large study, Pocock et al. (1984) evaluated the relationships between blood Pb concentrations, hypertension, and renal function indicators in a clinical survey of 7,735 middle-aged men from 24 British towns. The associations between systolic blood pressure and blood Pb levels, although small, were statistically significant ($p < 0.01$). Analyses of data in men categorized according to blood Pb concentrations indicated increases in blood pressure only at lower blood Pb concentrations; no further significant increments in blood pressure were observed at higher blood Pb. Of those men with blood lead over 37 μg/dL, 30% had hypertension, compared with 21% of all other men combined ($p < 0.08$). Pocock et al. (1984) interpreted their findings as suggestive of increased hypertension at blood Pb over 37 μg/dL, but not at the lower concentrations typically found in British men. However, more recent analyses of the same data reported by Pocock et al. (1985) indicated highly statistically significant associations between both systolic ($p < 0.003$) and diastolic ($p < 0.001$) blood pressure and blood Pb, when adjustments were made for variation due to site (town) in multiple regression analyses.

An ideal opportunity to separate the role of Pb from a wide range of potentially confounding nutritional factors was presented by the large data set from the second National Health and Nutrition Examination Survey of 1976–1980 (NHANES II), based on a random stratified sample of the U.S. population. Pirkle et al. (1985) described the results of their data analyses of 40- to 59-year-old white men from the survey population. After adjustment for age, body-mass index, all measured nutritional factors, and blood biochemistry factors were adjusted for a multiple linear regression model, the relations of both systolic and diastolic blood pressures to blood Pb were statistically significant ($p < 0.01$). In an examination of the NHANES II population in the age range from 12–74, Harlan et al. (1985) also found a direct relation between blood Pb and systolic and diastolic blood pressures.

In the Pirkle et al. (1985) analyses, the authors incorporated additional variables, with particular attention directed to the stability and significance of the Pb coefficient in the presence of nutritional factors and blood biochemistry. Their objective was to estimate conservatively the strength and independence of the relationship between blood pressure and blood Pb. Therefore, to provide an unusually rigorous test of the independent significance of blood Pb, 87 nutritional and biochemical variables in NHANES II were included in the stepwise regression. In addition, to account for possible curvilinear relations, squared and natural logarithmic transformations of almost all these variables were also included (see Table B-1).

TABLE B-1 Variables Included in the Stepwise Regression Analysis, White Males Aged 40–59 years, NHANES II, 1976–1980[a]

Age[b]	Dietary vitamin A[c]
Age-squared[b]	Dietary vitamin C[c]
Body mass index	Dietary thiamine[c]
Dietary sodium[c]	Dietary riboflavin[c]
Salt-shaker sodium	Dietary niacin
Dietary sodium × salt-shaker sodium	Serum cholesterol[c]
Dietary potassium[c]	Serum vitamin C[c]
Dietary sodium-potassium ratio	Serum iron[c]
Dietary calcium[c]	Serum transferrin saturation
Dietary phosphorus[c]	Serum zinc
Dietary protein[c]	Serum copper[c]
Dietary fat[c]	Serum albumin[c]
Dietary carbohydrate[c]	Hemoglobin[c]
Dietary cholesterol[c]	Red blood cell count
Dietary saturated fatty acids[c]	Ethanol consumption per week[c]
Dietary oleic acid[c]	Cigarettes smoked per day
Dietary linoleic acid[c]	Total dietary grams[c]
Dietary iron[c]	Total dietary calories[c]

[a] Reprinted with permission from Pirkle et al., 1985.
[b] This variable was forced into each regression to remove any possible age effect on blood pressure.
[c] The natural log and squared transformation of these variables were also included in the stepwise regression.

When they included the nutritional variables, the blood analytes, and their curvilinear transformations, Pb remained significantly associated ($p < 0.01$) with both systolic and diastolic blood pressures. Furthermore, segmented regression analyses indicated that there was not a threshold blood Pb content below which Pb was not significantly related to blood pressure.

Between 1976 and 1980, the mean blood Pb levels in the NHANES II population dropped by 37% owing to reductions in the amount of Pb used in gasoline. This much reduction in blood Pb in this population would be expected to result in a 17.5% decrease in diastolic blood pressure of at least 90 mm Hg, a value used to define hypertension.

Considering the relatively unusual nature of the blood Pb–blood-pressure relation (i.e., characterized by large initial increments in blood pressure at relatively low blood Pb concentrations, followed by leveling off of blood-pressure increments at high blood Pb concentrations), it is not surprising that it was not anticipated by results of animal studies. Many animal studies emphasize results from exposures at higher doses, where results tend to be more definitive. Yet, in retrospect, the human results were consistent with biphasic blood-pressure increases observed in response to blood Pb increases in the rat (Victery et al., 1982a,b). The unusual exposure-response relation might also account for the failure of earlier human studies to find consistent relations between blood pressure and blood Pb in study groups with mild to moderate increases in blood lead.

In summary, the use of a very large set of high-quality data covering a wide range of possibly confounding variables allowed a clear-cut determination of the effects of blood Pb on blood pressure for a relatively low range of blood Pb concentrations (5–35 μg/dl). The studies also demonstrated the utility of comprehensive data sets from representative populations for determining the effects of specific components in environmental exposures.

RADON DAUGHTERS AND CIGARETTE SMOKE

This case study illustrates the interaction of two pulmonary carcinogens, cigarette smoke and radon daughters, to produce an increase in the total incidence of bronchogenic carcinoma in humans. Both of these agents are inhaled as gaseous suspensions and are delivered to the epithelium. The result of combined exposures is to increase the number of carcinomas, and this increase is interpreted as additive by several investigators, whereas others consider the effect as multiplicative. Exposure to radon daughters and cigarette smoke also appears to decrease the latency induction period for lung cancer. Animal studies in general have supported the human epidemiologic and pathologic findings that the combined exposure to radon daughters and cigarette smoke yields a greater number of malignancies than that observed for either carcinogen alone.

Several of the cigarette-smoke–radon-daughter studies required the translation of the natural circumstances and organ systems to demonstrate the increased incidence of tumors. For example, McGregor (1976) used cigarette smoke condensate rather than native smoke. In addition, whereas alpha particles are generated by radon-daughter decay, this study used beta particles. Finally, the test organ was not the lung, but rat skin. These changes in the nature of the carcinogenic agents, in the concentration and manner in which they were applied, and in the target organ itself make this example questionable as an analogue of the human events.

The studies of Cross et al. (1982), in our opinion, have been incorrectly interpreted and are also of questionable utility. These investigators found more lung cancers in nonsmoking radon-daughter-exposed dogs than in a group exposed to cigarette smoke. These findings were interpreted as "protective" effect. Such conclusions are unwarranted, because dogs do not inhale cigarette smoke in a manner or quantity analogous to those of humans.

In fact, a major problem in virtually all animal studies in which inhalation of cigarette smoke is a requisite part of the design is the delivery of the smoke to the target organ in concentrations, temperature, age, and quantity identical to that inhaled by people. For one thing, there is a difference in inhalation pathway (i.e., nasal versus oral). In addition, most mammals find the inhalation of cigarette smoke so irritating that they must be forced to inhale. The volume of diluent air, the quantity of the cigarette smoke received, and the depth and frequency of inhalation are nonuniform and unlike the breathing pattern exhibited by human smokers. Furthermore, human lung diseases associated with cigarette smoke inhalation require heavy exposures (one or more packs per day) for periods of 20 or more years. Such exposures for such a duration are not feasible in animal studies. Thus, exposures of animals to cigarette smoke by inhalation are not adequate to initiate the disease states associated with human smoking.

Radon is an inert gas that results from the radioactive decay of radium-226; it has a half-life of 3.8 days. Other decay products resulting from the sequence of radioactive decay are polonium-218, lead-214, bismuth-214, and polonium-210. These elements are known collectively as radon daughters and have a collective half-life of 30 minutes. Small amounts of radon are ubiquitous, can be found in all rocks and soil, and are constantly emanating from ground level and diffusing into the atmosphere. Radon concentrations are increased wherever closed spaces exist around rock or soil and where phosphates or uranium ores or ore tailings are deposited. When radon decays, its daughter atoms, being heavy metals and ionized, attach to adjacent objects, surfaces, or dust.

When inhaled into the respiratory tract, radon decays rapidly. In so doing, it produces irradiation of the adjacent cells before it can be removed by normal clearance mechanisms. Of the radiation dose, 95% is delivered to the epithelium as alpha radiation; the beta and gamma energies are generally insignificant and diffusely distributed. Exposure is measured in working levels (WL).

One WL is the concentration of any combination of short-lived radon daughters in air whose complete decay produces 1.3×10^5 MeV of alpha energy per liter. Exposure at this level for 1 working month, about 170 hours, constitutes an exposure of 1 working level month (WLM). The maximal permissible exposure for underground miners is 4 WLM/year, an equivalent of 0.3 WL/month. Levels in mines have cumulatively reached as high as 10,000 WLM. Average levels in homes appear to be about 0.004 WL in the United States, although some homes have been found to have levels as high as 10 WL.

When the first bronchogenic carcinoma was observed in a nonsmoking uranium miner in 1965 (described by Archer, 1985), there had already been 80 lung-cancer deaths in cigarette-smoking U.S. uranium miners. Because of the increased expectancy of lung neoplasms in cigarette smokers, it was quite natural to assume that the cause of all lung cancers in this group of miners was cigarette-smoking. In 1969, a second primary bronchogenic carcinoma was observed in a nonsmoking uranium miner (Archer, 1985). This case represented a significant excess of lung cancers from what might be expected in this nonsmoking population.

As the mortality study of uranium miners continued, Archer (1985) noted that nonsmoking uranium miners experienced 7 times the incidence of lung cancer observed in nonsmoking persons who were not miners, and smoking uranium miners had 9.5 times the incidence of lung cancers found in nonminers with similar smoking histories. Despite the fact that uranium miners smoked more than the comparably aged nonminer U.S. males, the excess exposure to cigarette smoke explained only a small part of the observed increase in lung cancer. This unexplained increase in cancer incidence suggested an interaction between cigarette-smoking and radiation exposure in the miners.

Archer (1985) described additional studies which showed that the lung-cancer rate among smoking uranium miners declined after age 65. The explanation suggested was that the latency induction period had been shortened in smoking miners. Interpretation of this finding was that cigarette smoke might act as a promoter, whereas radiation acted as the initiator.

Axelson and Sundell (1978) suggested that the increased incidence of lung cancer reflected an additive effect between smoking and radon-daughter exposure. Damber and Larsson (1985) interpreted their findings as a multiplicative effect. The studies of Hornung et al. (1981), Edling (1982), and Radford and St. Clair Renard (1984) all supported an additive effect, rather than a multiplicative one.

The early studies of the cell type of bronchogenic carcinoma seen in uranium miners suggested that there were greater numbers of undifferentiated small-cell carcinomas in this group than in the lung cancers of nonminers. The increased proportion of small-cell or differentiated lung cancers, which appeared to be the same in smoking and nonsmoking uranium miners, suggests that mining exposure and radiation were responsible for any difference in the number of undifferentiated carcinomas. Recent pathologic studies have not confirmed this increase in undifferentiated small-cell cancers in miners. All cell types—the squamous carcinoma, the adenocarcinoma, and the small-cell undifferentiated carcinoma—have been observed in cigarette-smoking uranium miners in numbers similar to that present in cigarette smokers who are not miners.

Factors that might influence the cell type and progression of lung cancers in U.S. uranium miners are age at the start of mining, high radiation exposure rates, and cigarette-smoking. The risk of lung cancer appears to decline in miners who have reached age 65 and have had 25 or more years of latency.

Several animal studies have addressed the interaction of cigarette smoke, cigarette-smoke condensate (CSC), and ionizing radiation. McGregor (1976) studied the interaction of CSC and beta particles on rat skin and observed an increase in skin tumors 3 times greater than the increase produced by radiation alone. CSC alone did not produce tumors. Treatment with both agents caused tumors to appear earlier than in the radiation-only group. Nenot (1977) used cigarette-smoke inhalation and americium-241 applied to the lungs in rats. The animals receiving the combination of smoke and radiation had an increased yield of lung cancers, which appeared earlier than with radiation alone.

In summary, epidemiologic studies of exposure to radon daughters and cigarette smoke have demonstrated two effects. The first is an additive effect of the two agents on the number of cancers induced. The second is the decrease in induction latency period, which produces a different time distribution of tumors in smokers and nonsmokers: the tumors in smokers appear earlier. We have seen comparable results in animals exposed to components of cigarette smoke at high concentrations, but not when the animals were exposed to fresh smoke by inhalation. The discordance in the latter case is most likely due to technical limitations in our ability to produce conditions that match human smoking.

ASBESTOS EXPOSURE AND CIGARETTE-SMOKING

This case study is one of the most current and well-recognized examples of how two distinct agents administered together can produce an increased incidence of lung cancer that is greater than that predicted from the administration of either agent alone. The increase in lung-cancer incidence in cigarette-smoking asbestos workers is considered multiplicative by most investigators who have studied the problem. That is in contrast with the increased incidence of lung cancer in uranium miners, in whom it has been observed that the increase in lung-cancer incidence from the combined exposure is additive.

This case study also differs from that of the radon-daughter–cigarette-smoke example, in that asbestos, unlike radon gas, is a fibrous mineral particulate material that enters the lung via the airways and remains in the lung for long periods. There is ample evidence that asbestos alone is a mild carcinogen. However, the amount of asbestos required to initiate carcinogenesis is not known. Although cigarette-smoking unquestionably increases the incidence of primary epithelial lung neoplasm, it has no apparent effect on the malignant tumors arising in the pleural, peritoneal, or pericardial mesothelium. Mesotheliomas, as these malignant tumors are called, have a definite relation to inhalation of asbestos fibers. However, mesotheliomas have occurred after relatively short and light asbestos exposures, as well as after prolonged and heavy occupational exposures. The dose-response relation is therefore variable, and the threshold dose required for mesotheliomas to occur might be exceedingly small or nonexistent.

It is generally accepted that the extent and severity of asbestosis, the disease state in which scarring or fibrosis develops, is related to the duration and quantity of asbestos to which the person has been exposed. There is no evidence, however, that cigarette-smoking accelerates or modifies the extent of fibrosis produced by asbestos exposure. Animal studies, when properly designed, have demonstrated the production of primary lung cancer, pleural mesotheliomas, and asbestosis. Studies in which combined exposures of animals to cigarette smoke and asbestos fibers were used were difficult to interpret, because the number of lung tumors that occurred as a result of combined exposure was no greater than that produced by asbestos exposure alone. It is likely that the design of that part of the study, which required the delivery of cigarette smoke to the lungs of the animal subjects, was unreliable because of the difficulty, previously noted, in having animals inhale cigarette smoke in the manner and quantities of human inhalation. However, when carcinogenic agents present in cigarette smoke, such as benzo[a]pyrene, are administered with asbestos in a syngeneic tracheal graft or organ culture system, the increase in carcinomas in the target tissue is greater than that initiated by either agent separately.

In 1935, Lynch and Smith reported the first case of lung carcinoma associated with asbestosis and silicosis. Sporadic case reports of bronchogenic carcinoma in asbestos-exposed workers appeared in the medical literature between 1935 and 1955. In a landmark 1955 study, Doll reported increased mortality from lung cancer in British asbestos workers. That observation was confirmed by Selikoff et al. in 1968. The latter researchers also evaluated and correlated smoking habits with the incidence of lung cancer in asbestos-insulation workers. In the study, no lung cancer was detected in the 48 workers who never smoked, nor in 39 workers who smoked only pipes or cigars. Of the remaining 283 cigarette-smoking workers, 24 died of lung cancer in the followup period, whereas only 2.98 deaths from lung cancer were expected on the basis of statistics in smokers.

In 1979, Hammond et al. expanded the evaluation of the role of cigarette-smoking in the development of primary lung cancer among asbestos-insulation workers. Of the 1,332 deaths in that group, 314, or 23.6%, were from lung cancer. At 20 years after onset of asbestos exposure, the death rate of nonsmoking controls was 11.3 per 100,000, whereas the nonsmoking asbestos workers had a death rate of 58.4 per 100,000 and thus a mortality ratio 5.17 times greater. The death rate of smokers in the control group (no asbestos exposure) was 122.6 per 100,000, whereas the smoking asbestos workers had a death rate of 601.6 per 100,000 and thus a mortality ratio 4.9 times greater. The lung-cancer risk for the asbestos-exposed smoking group was 53.24 times that of the nonsmoking, non-asbestos-exposed smoking group, a figure that reflects a multiplicative relation between the two variables (5.17 and 10.85, see Table B-2). The mortality ratio for lung cancer in pipe and cigar smokers was 7.02. There appeared to be a dose-response relation between the amount of cigarette-smoking and the asbestos exposure. In the same study, ex-smokers with asbestos exposure had a mortality ratio of 36.56 over those who smoked less than one pack per day, 50.82, whereas smokers of a pack or more per day had a ratio of 87.36.

Similar findings were reported by Berry et al. in 1972. This group studied 1,203 male asbestos workers, 74.5% of whom were cigarette smokers. A dose-response relation of cigarette-smoking with asbestos exposure was observed in both men and women: with greater exposure, the ratio of observed to expected lung-cancer deaths increased. McDonald and his colleagues (1980) further confirmed this effect on Canadian miners and millers, as did Meurman and colleagues (1974) in Finnish anthophyllite miners and millers.

The carcinogenic effect of asbestos-fiber inhalation without cigarette-smoking has been demonstrated in several independent studies. In addition to the studies of Selikoff et al. (1968), those of Berry et al. (1972), McDonald et al. (1980), Meurman et al. (1974), and Liddell et al. (1982) reported deaths from lung cancer among nonsmoking asbestos-exposed workers. Studies have not established a threshold quantity of asbestos required to increase the risk of lung

TABLE B-2 Age-Standardized Lung Cancer Death Rates[a] for Cigarette Smoking and/or Occupational Exposure to Asbestos Dust Compared with No Smoking and No Occupational Exposure to Asbestos Dust[b]

Group	Exposure to Asbestos?	History Cigarette Smoking?	Death Rate	Mortality Difference	Mortality Ratio
Control	No	No	11.3	0.0	1.00
Asbestos workers	Yes	No	58.4	+47.1	5.17
Control	No	Yes	122.6	+111.3	10.85
Asbestos workers	Yes	Yes	601.6	+590.3	53.24

[a] Rate per 100,000 man-years standardized for age on the distribution of the man-years of all the asbestos workers. Number of lung cancer deaths based on death certificate information.
[b] From Hammond et al. (1979).

cancer, because the asbestos exposure has not been quantified. Furthermore, as the dose of asbestos decreases, mortality statistics become less accurate.

The risk of developing lung cancer after cessation of cigarette-smoking in asbestos-exposed workers appears to decrease in a manner similar to that seen in non-asbestos-exposed cigarette smokers who stop smoking. These findings have been observed by Hammond et al. (1979) and independently by Walker (1984). Seidman and colleagues (1979) studied workers who received brief but intense exposures to asbestos and were studied 35 years later. The lung-cancer mortality ratio was increased and appeared to remain increased for the entire period after exposure. An occupational exposure to amosite asbestos for as short a time as 1 month was associated with increased lung-cancer mortality. The likelihood of cancer appeared to increase with the duration of exposure after a latent period of at least 25 years. The length of the latent period also appeared to vary with the extent and duration of the exposure. Saracci (1977) reviewed the reports relating cigarette-smoking and asbestos exposure from five major studies and concluded that the relation was consistent with a multiplicative model.

Rats have been exposed to asbestos fibers by inhalation in several carefully performed studies (Davis et al., 1980; Wagner et al. 1980). After exposures of a year or more, approximately 25% of the animals developed primary lung carcinomas, which were most commonly adenocarcinomas. All three major types of asbestos produced these carcinomas, but there was a greater accumulation of amphiboles than of chrysotiles in the lung.

Wehner et al. (1975) reported that hamsters exposed to cigarette smoke and asbestos by inhalation developed adenomas, papillomas, and adenocarcinomas. The results of the study are difficult to interpret, because the number of tumors that developed in the group receiving the combined cigarette-smoke and asbestos exposure was no greater than that produced by asbestos inhalation alone. Because difficulties in delivery of cigarette smoke to the lung of small mammals undermine our assurance that the animals received the smoke, several investigators have studied the effect of asbestos in association with benzo[a]pyrene (BaP), a polycyclic aromatic hydrocarbon present in cigarette smoke. A striking increase in the number of benign and malignant neoplasms of the lung and airway was seen as a result of this combined exposure, compared with the number seen after exposures to asbestos or BaP alone. These findings support the observations in humans of an increased effect from the combination of asbestos and cigarette smoke, compared with either agent alone.

Topping and Nettesheim (1980) used an unusual syngeneic tracheal graft system to study the effects of asbestos and polycyclic hydrocarbons. Carcinomas occurred when the grafts were treated with dimethylbenzanthracene (DMB) first and later with chrysotile asbestos. The DMB alone did not cause tumors—an indication that asbestos had acted as a promoting agent. Mossman and Craighead (1982) reported similar findings in an organ-culture system of hamster tracheas. Carcinomas developed when 3-methylcholanthrene (3MC) was used to coat the surface of crocidolite fibers that were then placed on the epithelial surface of the tracheal organ culture. Asbestos was considered the carrier of the 3MC, because other minerable dusts, such as kaolin and carbon, also caused tumor formation if coated with 3MC.

Although the evidence that epithelial malignant tumors develop in increased numbers as a result of combined exposure to cigarette smoke and asbestos inhalation is compelling, that is not the case for malignant mesotheliomas. Mesotheliomas are rare tumors that are believed to arise from the lining cells of the serous cavities, namely the pleura, peritoneum, and pericardium. Their association with asbestos exposure was first documented by Wagner et al. (1960) and has been confirmed by Selikoff et al. (1968), Newhouse et al. (1972), McDonald et al. (1980), and others. Cigarette-smoking does not appear to influence the incidence, latency, or progression of these tumors.

The combined exposure to asbestos and cigarette smoke in humans also has effects on functional responses of the lungs and airways that are modifications of the effects of either agent acting alone. Prolonged exposure to cigarette smoke alone is characterized in humans by airway obstruction, parenchymal destruction with little or no fibrosis, and increased lung volume. There is also often increased production of mucus associated with mucous gland hypertrophy. However, asbestos inhalation in humans over long periods leads to parenchymal destruction with fibrosis, reduced lung volumes, and minimal airway narrowing. In asbestos-exposed heavy cigarette-smokers, the predominant functional changes are most commonly airway obstruction and reduced lung volume. This combination of functional changes reflects the combined effects of both agents.

Regan and coauthors (1971) evaluated the discriminatory power of various pulmonary function tests to distinguish the effects of smoking from those of asbestos exposure. Decreases in diffusing capacity for carbon monoxide and decreases in vital capacity were the best measures of severity of obstructive and restrictive lung disease in persons with combined exposures, but these tests do not clearly distinguish between the two agents. A good indicator to distinguish the effect of cigarette-smoking is the 1-second forced expiratory volume (FEV_1) as a percentage of forced vital capacity (FVC). The test reflects airflow obstruction in large or proximal airways; however, it is not a good index of small-airway function. Wright and Churg (1984) have reported that asbestos produces a much greater change in membranous bronchioles than does cigarette-smoking alone. But the extent of the lesions produced by asbestos and cigarette-smoking, compared with those produced by asbestos alone, was not clearly distinguished. Begin et al. (1983) evaluated small airways in cigarette-smoking asbestos miners and millers in Quebec by physiologic and lung biopsy studies. They found that nonsmokers had relatively normal function, whereas smoking miners and millers had clear evidence of small-airway obstruction and a 3- to 4-fold increase in upstream resistance at low lung volumes. The small-airway dysfunctions observed in asbestos workers who are not cigarette smokers were not of sufficient severity to cause alterations in the FEV_1/FVC ratio.

Persons with a long history of cigarette-smoking and asbestos exposure have a higher prevalence of radiologic abnormalities characterized as interstitial fibrosis than do nonsmoking asbestos-exposed persons. The differences are not apparent in populations with a higher prevalence of roentgenographic fibrosis and presumably higher asbestos exposures. The study of Harries et al. (1972) suggests that a shorter asbestos exposure might be required to produce an abnormal chest roentgenogram in smokers than in nonsmokers. There is no other evidence to suggest that smokers develop more severe fibrosis than nonsmokers.

A number of immunologic changes have been observed in asbestos-exposed workers, including humoral and cellular immune responses. Lange (1980) reported that a proportion of asbestos-exposed workers with roentgenographically demonstrable asbestosis had increased circulating immunoglobins A and G. Those findings were believed to be due to asbestos exposure, rather than to cigarette-smoking. Wagner and colleagues (1960) studied the effect of smoking and asbestos exposure on T-lymphocyte number and subset formation. Persons with roentgenographic asbestosis who smoked had a greater number of T helper cells than those who did not smoke, whereas suppressor T cells were not affected. Wagner (1980) reported that asbestos exposure was associated with a decrease in number of total T cells, a decrease in suppressor T cells, and an increase in the helper subset. Smoking history did not influence those results. In other studies, heavy smoking was found to cause an increase in total T lymphocytes, a decrease in the proportion of T helper cells, and an increase in T suppressor cells; this creates a decreased ratio of helper to suppressor cells. DeShazo and colleagues (1983) observed that asbestos workers had significantly increased proportions and total numbers of B and T lymphocytes in peripheral blood and a proportionate decrease in helper T cells. Those changes had no relation to radiologic categories of pneumoconiosis, cumulative asbestos exposure, or pulmonary functional abnormality. There is apparently a lack of uniform results concerning the effect of asbestos and cigarette-smoking on T-lymphocyte number and activity. More work is required to clarify this problem.

CIGARETTE-SMOKING AND ALCOHOLIC-BEVERAGE CONSUMPTION

Both cigarette-smoking and alcoholic-beverage consumption are risk factors for cancers of the oral cavity. Because these exposures commonly occur together, one can usefully ask whether concurrent exposure to both increases the risk above that expected from simply summing the effects of either exposure alone. That was done in a useful paper by Rothman and Keller (1972), summarized here.

Before examining the effects of joint exposure, we must define the measurements that quantify the population response to individual exposures and define how joint exposures may be classified. Disease risk may be expressed as time-conditioned probability of disease occurrence (e.g., 10 cases per 100,000 persons per year). We can assume that the first of two independently acting exposures increases risk by an amount, x (i.e., the disease rate is higher in the exposed group than in the unexposed by x). Let us assume that a second exposure increases risk by an amount, y. If the result of both exposures is to increase risk by $x + y$, the exposures are said to act *independently*. If the risk resulting from both exposures is greater than $x + y$, the interaction is said to be *synergistic*; if the risk is less than the sum of the two acting alone, the interaction is defined as *antagonistic*.

The relative risk (risk ratio) is commonly used in epidemiologic studies as a measure of risk. It is defined as the probability of disease in an exposed group divided by disease probability among an unexposed population that is otherwise similar (with respect to age, race distribution, socioeconomic status, etc.). If unexposed persons have a disease rate R_0 and exposed persons have a rate R_1, the relative risk is R_1/R_0. If another exposure increases the baseline risk R_0 by y to a level $R_0 + y$, then independence of action of the two exposure factors implies that the relative risk due to exposure x would be lower in the presence of the other risk factor than in its absence. In general, the proportional or relative increase in risk due to a constant addition to the disease rate among exposed persons decreases with the magnitude of the baseline risk. Thus, a decrease in relative risk from one risk factor in the presence of another is consistent with a simple additive risk model.

Rothman and Keller (1972) used data from a study of mouth and pharynx cancers (Keller and Terris, 1965) to examine whether cigarette-smoking and consumption of alcoholic beverages are independent (i.e., simply additive), synergistic, or antagonistic. After excluding people with incomplete smoking or drinking histories, the study included 483 cases and 447 controls. The cases in males were diagnosed with squamous cell carcinoma of the mouth or pharynx. Controls were matched to cases by age and sex. Histories of smoking and drinking, routinely taken by the admitting physician, were abstracted from the clinical record. Smoking was expressed as number of cigarettes per day, considering one cigar the equivalent of four cigarettes and one pipeful of tobacco the equivalent of two cigarettes. Alcohol was expressed as ounces of alcohol per day. Although stratified by age, relative risks were calculated for various combinations of smoking and drinking behavior.

When smoking and drinking are considered in a simple table dichotomous

TABLE B-3 Relative Risks of Oral Cancer According to Presence or Absence of Two Exposures—Smoking and Alcohol Consumption[a]

Alcohol Consumption	Smoking	Relative Risk
No	No	1.00
No	Yes	1.53
Yes	No	1.23
Yes	Yes	5.71

[a] Data from Rothman and Keller (1972).

for both exposures, there is a strong suggestion of synergy (Table B-3). In the absence of drinking, smoking adds 0.53 unit of risk to the baseline risk of 1.00. In the absence of smoking, drinking adds 0.23 unit of risk to the baseline. If each of these simple dichotomous exposures acted independently, the contributions to risk would be additive, and the net contribution of both exposures would be to increase risk by 0.53 + 0.23 for an excess risk of 0.76, instead of the observed excess risk of 4.71.

The detailed data were used by Rothman and Keller (1972) to examine risk more carefully. Table B-3 shows the relative risk of oral cancer according to level of exposure to smoking and alcohol. In Table B-3, the result of increasing each exposure is observed at every level of the other; that illustrates the individual effects of the exposures. In addition, the observed risks in the lower right portion of Table B-3 are greater than expected on the basis of simple additivity. The expected relative excess risk in the highest smoking and drinking stratum is 1.33 + 1.42 = 2.76, far less than the observed level of 14.5.

Caution is warranted in interpreting these results, however. Most of the excess risks that suggest synergy occur in the cells along the bottom row and right-hand column of the table. These strata represent open-ended exposure categories, where an excess of one exposure might account for the excess risk observed, in the absence of synergy. It would not be correct, for example, to

TABLE B-4 Relative Risks of Oral Cancer According to Level of Exposure to Smoking and Alcohol[a]

Alcohol, oz/day	Smoking (cigarette equivalents/day)			
	0	<20	20–39	≥40
0	1.00	1.52	1.43	2.43
<0.4	1.40	1.67	3.18	3.25
0.4–1.5	1.60	4.36	4.46	8.21
≥1.6	2.33	4.13	9.59	15.5

[a] Adapted from Rothman and Keller (1972).

assume that nonsmokers who consumed 1.6 ounces per day consumed the same amount of alcohol as those smoking 20–39 cigarettes per day and drinking ≥ 1.6 ounces of alcohol per day. The exposures are correlated and the categories open-ended with respect to alcohol ingestion. Observed risks in cells adjacent to those in the lower right-hand corner of Table B-3 also exceed expected values and offer stronger evidence of synergy. But the suggested interaction might have arisen because the categories are too broad. Moreover, a randomly low value in one of the two cells used to generate the expected values can falsely suggest synergy. Rothman and Keller (1972) concluded that the data in Table B-4 suggest a combined effect equal to the sum of two strong individual effects plus a synergistic component. They believed that the evidence from this single study was not strong enough to warrant completely discarding the simple model of independent effects.

Data from another case-control study of oral cancer by Wynder et al. (1957) were transformed by Rothman and Keller and also presented in tabular form (Table B-5). A different classification scheme was used to produce more stable estimates of relative risk, but evaluation of synergy should not depend on the classification used. They concluded that the results shown in Table B-5 suggest that action of the two exposures is additive and independent.

TRIHALOMETHANES AND OTHER BYPRODUCTS OF CHLORINATION IN DRINKING WATER

This case study demonstrates how epidemiologic methods have been used to evaluate human risk from exposure to complex mixtures of chlorination byproducts in disinfected drinking water. It is generally accepted that the practice of chlorine disinfection, used since 1909 in the United States (Johnson, 1911), has been an important factor in reducing morbidity and mortality from various waterborne pathogens, although numerical estimates are not available. A

TABLE B-5 Relative Risk[a] of Oral Cancer According to Level of Exposure to Smoking and Alcohol[b]

Alcohol, units/day[c]	Smoking (cigarettes/day)			
	<15	16–20	21–34	≥35
<1	1.00	2.86	1.79	8.40
1–2	1.70	2.05	1.94	3.88
3–6	6.20	7.02	8.91	5.33
>6	9.69	11.6	17.0	19.4

[a] Risks expressed relative to risk of 1.00 for persons who smoked fewer than 16 cigarettes per day and drank alcohol less than 1 unit per day.
[b] Unadjusted for age. Adapted from Rothman and Keller (1972).
[c] One unit of alcohol equals 1 oz. of whiskey or 8 oz. of beer.

suggestion that the time-tested benefits of chlorination can be partially offset by an increase in the burden of chronic disease from chlorination byproducts was raised in 1974 and 1975. Rook (1974) and Bellar et al. (1974), working independently, showed that chloroform and other trihalomethanes (THMs) are created during chlorine disinfection. Environmental surveys demonstrated that concentrations are much higher in treated surface waters than in water from underground sources. Shortly thereafter, a feeding study of chloroform demonstrated its carcinogenicity in rodents (Page and Saffiotti, 1976). Later work showed that THMs are accompanied by a complex mixture of higher-molecular-weight nonvolatile chlorinated organic substances (Stevens et al., 1985). Specific mutagens, including chlorinated aldehydes and ketones, have been identified (Meier et al., 1985). In most treated waters, more than half the organically bound chlorine is associated with this nonvolatile fraction (Stevens et al., 1985).

Assessment of possible links of chlorination byproducts with cancer in human populations started soon after the discovery of THMs in treated water and continues today. A central issue in epidemiologic studies of chlorination byproducts, as well as other complex mixtures, is how best to define exposure. Ideally, the measure used would provide accurate estimates of past exposure to the biologically active factors in the mixture.

Most epidemiologic studies of chlorination byproducts have relied on the large differences in contaminant concentrations between water from surface and ground sources and between chlorinated and nonchlorinated water, to derive relatively simple, categorical definitions of exposure. A few studies have used recent measurements of THMs to define likely past concentrations. The use of recent THM measurements to impute past exposures, although satisfying a need for quantitative estimates, might be misleading. THM contents from any supply show large diurnal and seasonal variations. THM estimates from a single measurement, therefore, can provide but a crude estimate of past average exposures. The biologic activity of interest might reside not in the easily measured THMs, but rather in the higher-molecular-weight fraction of the chlorination byproduct mixture. The degree of correlation between THMs and biologically active nonvolatile compounds, although assumed to be high, is not well documented in field measurements and might be variable. In the face of these difficulties, comparisons of risk between groups or individuals exposed to chlorinated surface water, as opposed to persons who used groundwater, could be as valid and useful as comparisons based on actual THM measurements.

Just as toxicologic assessment of chlorination byproducts and other chemicals in drinking water have evolved from crude evaluations of complex mixtures to identification of individual chemicals, so epidemiologic studies have progressed from broad assessments, with little ability to distinguish specific exposures or control for potential confounders, to more refined "analytic" studies. The first epidemiologic studies were conducted shortly after information on specific chemical contaminants became available. Most studies were ecologic in design, comparing the geographic distribution of site-specific, age-adjusted cancer mortality rates among U.S. counties with the distribution of water-supply characteristics or measured concentrations of drinking-water contaminants. One of the earliest studies (De Rouen and Diem, 1977) compared cancer rates in Louisiana parishes (counties) served by Mississippi River water (with a large number of known organic contaminants) with cancer rates in parishes served by other, mostly groundwater, sources. A multiple-regression model was used in which the independent predictor variables included the percentage of the parish population drinking water from the Mississippi, parish urbanicity, median income, and several parish-level industrial variables. Dependent variables in separate regression models were average annual age-adjusted mortality rates for 1950–1969 for cancers of the gastrointestinal and urinary tracts and total cancer. Statistical tests evaluated the magnitude and statistical significance of correlations of the cancer rate with the drinking-water variable, after other parish characteristics were adjusted for.

For the development of exposure indexes, other demographic studies of cancer mortality and water quality used data from a 1963 U.S. Public Health Service Inventory of Municipal Water Supplies (1964) to calculate the percentage of county populations served by surface or ground sources or by chlorinated or nonchlorinated supplies. Salg (1977) examined the association between water-quality factors and cancer mortality in the 346 counties of the Ohio River drainage basin, using as exposure variables the percentage of each county's population served by surface water. Another study (Kuzma et al., 1977) looked for associations of water quality and cancer rates in Ohio counties, using a dichotomous exposure variable that indicated whether surface water or groundwater was used by a majority of a county's population. Other studies (e.g., Burke et al., 1983) used as exposures THM contents in the predominant county water supply, according to U.S. Environmental Protection Agency surveys.

Reviews of the ecologic studies of cancer and drinking-water contaminants (e.g., Craun, 1985) noted that, although there were many inconsistencies in their results, three cancer sites—bladder, colon, and rectum—appeared to be associated with one or another water quality variable more often than might have occurred by chance. It was suggested that these three sites deserved further evaluation in more highly focused case-control studies.

The first case-control studies selected deceased cases and controls from computerized listings of state vital-statistics bureaus. Addresses on death certificates served as links to information on drinking-water source and treatment from the community water supply at the time of death. With this information, the most recent drinking-water source and treatment were characterized for each study subject, and the frequency of use of different types of water sources among cases was compared with that of controls. Although these studies represented a major step forward, in that exposure and disease information was on an individual, not group, level, they shared some of the weaknesses of indirect, ecologic studies. Among the weaknesses are the potential for exposure misclassification due to changes in water supply, migration or other factors, and the inability to account for other risk factors that can confound relations with water contaminants or interact with them. That can be of special importance when relatively small effects are expected. Five studies (Alavanja et al., 1978; Kuzma et al., 1977; Brenniman et al., 1980; Gottlieb and Carr, 1982; Kanarek and Young, 1982) examined several cancer sites as related to water source at the decedent's last address. In a sixth study (Young and Kanarek, 1983), drinking water of decedents was defined for the last 20 years of life.

Slightly increased relative risks were observed for cancers of the colon, rectum, and bladder in five studies, associated with either surface source (versus ground) or chlorinated source (versus nonchlorinated). Risk ratios from the five studies were in the range 1.0–2.0. The sixth study, looking only at colon and rectal cancer in relation to exposures up to 20 years before death, found no association for either site.

Risk ratios below about 2.0 from epidemiologic studies are often subject to question, even if they are statistically significant. Associations of that magnitude can be due to confounding from risk factors that are not ascertained. Nevertheless, when exposures are widespread in a population, even small increases in relative risk can translate into large numbers of excess malignancies. In addition, if the positive associations from the case-control mortality studies reflected causal relationships, the magnitude of the risk might have been underestimated, because the net effect of migration would be to decrease the accuracy of exposure information. Those factors justified continued research into water contaminants and human cancer.

Epidemiologic evaluation has continued in the form of case-control studies of incident cancers. Such studies typically gather information on a number of risk factors directly from cancer subjects, controls, or their next of kin. That information permits careful control for potential confounding by other risk factors and increases the precision of risk estimates from water-related exposures, because more accurate exposure information is available. Preliminary results from a study of colon cancer in North Carolina (Cragle et al., 1985) show associations with exposure to chlorinated surface water, especially among older groups exposed for at least 15 years.

A case-control study of the incidence of colon cancer in Wisconsin (Kanarek and Young, 1982) showed no association of risk with several different measures of past THM intake. Dose differences between exposed and unexposed persons in Wisconsin might not have been as large as in North Carolina, because relatively uncontaminated surface waters were used for drinking-water supplies in Wisconsin and had lower concentrations of chlorination byproducts.

A large case-control interview study (Cantor et al., 1985) of bladder cancer in 10 areas of the United States revealed increasing risk with tap-water intake. The risk gradient with intake was largely restricted to long-term consumers of chlorinated surface water and was not found among consumers of nonchlorinated groundwater. Relative to long-term consumers of groundwater, bladder-cancer risk increased with duration of exposure to drinking water from treated surface sources, especially among high-quantity consumers. Although most relative risks for bladder cancer were below 2.0, the findings of increased risk were generally consistent among geographic areas and between the sexes, and statistical tests for trends with tap-water intake were highly significant.

Results from carcinogenicity testing in animal feeding studies are available for the four major THMs in drinking water (Balster and Borzelleca, 1982), but not the nonvolatile chlorination byproducts. Extensive testing of nonvolatile concentrates has demonstrated mutagenicity in a variety of in vitro testing systems, as well as transformation of mammalian cells in tissue culture. Identification and toxicologic testing (in vitro) of several compounds of the chlorination-byproducts mixture have been conducted and have raised general concerns, but more specific implications for adverse human health effects are unclear.

COKE-OVEN EMISSIONS

The following discussion of the toxic effects of coke-oven emissions is presented to demonstrate that animal toxicity testing of complex mixtures is a valid means of predicting human disease resulting from environmental exposure to these mixtures.

From a toxicologic point of view, coke-oven emissions and related agents, such as coal tar, are probably the most widely studied complex mixtures. There is an overwhelming body of scientific evidence that these substances are carcinogenic. Coal-tar derivatives were probably responsible for the first recorded observation of occupational cancer, made in 1775 by Percivall Pott, who noted an excess of scrotal cancer in London chimneysweeps. The first experimental demonstration of chemical carcinogenesis was made by Yamagiwa and Ichikawa in 1918, when they applied coal tar to the ears of rabbits. Numerous epidemiologic and animal studies have confirmed the carcinogenic properties of combustion and distillation products of coal. In the present discussion, the relevant question is: Given the animal data, can we predict human health effects of coke-oven emissions?

The most extensive demonstrations of human disease resulting from coke-oven emissions have come from mortality studies. These studies show significant increases in lung and genitourinary cancer mortality associated with exposure to coke-oven emissions. The first of these reports was made in 1936 by investigators in Japan and England (Kennaway and Kennaway, 1936) who were studying lung-cancer mortality among persons employed in coal carbonization and gasification processes. Later studies, conducted in the United States by Lloyd (1971) and others, have clearly demonstrated substantial increases in lung-cancer mortality among coke-oven workers. Similar studies have shown that coke-oven workers have a significant increase in mortality due to cancer of the genitourinary system (IARC, 1985).

It is clear that human exposure to coke-oven emissions results in lung and genitourinary cancer. The question is: Could these diseases have been predicted from available animal data? At first glance, the answer appears to be no, because of the designs of the animal experiments and the epidemiologic studies. Coke-oven emissions cannot be generated in an experimental setting. For that reason, coal tar derived from coking operations has been used for animal research. The chemical composition of coal tar is very similar to that of coke-oven emissions, so the effects of coal-tar exposures on animals are thought to approximate closely the effects of coke-oven emissions on humans.

For 70 years, investigators have been producing skin cancer in experimental animals by dermal application of coal tar or coal-tar extracts. The results are consistent with observations of increased scrotal cancer in chimneysweeps made two centuries ago, but are not confirmed by modern epidemiologic studies of coke-oven workers. There are two possible reasons for the discrepancy. First, modern coke-oven operations and practices of hygiene have decreased dermal exposures. Second, and more probable, the epidemiologic studies used mortality data, and skin cancer is seldom fatal. The point is that animal data will not predict human disease if means are not available for gathering human data. That applies not only to nonfatal malignancies, but also to other disease end points that do not show up on death certificates.

In contrast with skin cancer, lung and genitourinary cancers have been detected readily in human epidemiologic studies, but not until recently in experimental animal studies and even then not genitourinary cancers. In this case, the discrepancy was probably due to the design of the animal studies, which used dermal exposures. It was not until epidemiologists discovered the relation between coke-oven emissions and lung and genitourinary cancers that experimenters began extensive use of nondermal routes of exposure. Experiments confirmed the epidemiologists' findings in the case of lung cancer. A similar situation existed for nonmalignant respiratory diseases that were detected by epidemiologists in coke-oven workers. Those diseases have been overlooked by experimenters, possibly because of the emphasis placed on carcinogenic end points. Animal studies must be designed appropriately, if their results are to predict specific human diseases.

It is clear that, for coke-oven emissions, the correlation between animal and human data is far from perfect. When we view the evidence retrospectively, it is evident that the design of animal experiments might not have produced the data necessary to make predictions of specific human diseases before their discovery by epidemiologists. It might never be possible to know a priori precisely which experiments to do to determine specific human health effects of complex mixtures. Given the imperfect nature of animal data, can they be used to make decisions concerning health threats posed by complex mixtures? Judging from our experience with coke-oven emissions, the answer is clearly yes. Although there are substantial differences between the human and animal data, we conclude that the adverse effects of coal tar seen in animals are echoed by the effects of coke-oven emissions in humans.

The above discussion of coke-oven emissions demonstrates that data generated by animal toxicity testing of complex mixtures can be a reasonable predictor of human disease, if the mixture being tested in the laboratory is representative of mixtures in the human environment, if studies are properly designed to detect diseases, and if we are aware that the specific diseases and target organs in test animals can vary from those in humans.

COAL-MINE DUST

The studies of lung disease in coal miners have focused principally on chest x-ray alterations, pulmonary function deficits, chronic bronchitis, and pathologic alterations of lung tissue. Attempts have been made to relate those to each other, as well as to an exposure index (e.g., years spent underground in coal mines) and dose (mass of respirable coal dust per cubic meter of air times total hours exposed). An immense amount of data has been generated. The purpose of this example is to document the correlations, point out the difficulties involved, and explain how animal experiments have or have not assisted in explaining some of the biologic response variables.

Coal workers' pneumoconiosis (CWP) was originally described in terms of descriptive pathology (Gough, 1947; Heppleston, 1947). The characteristic lesion is a focal collection of coal-dust-laden macrophages at the division of respiratory bronchioles that can exist within alveoli and extend into the peribronchiolar interstitium with associated reticulin deposits and focal emphysema (Kleinerman et al., 1979). The black lesions have been termed macules; they are 1–4 mm in diameter. Other lesions can occur in lungs of coal miners, including micronodules (< 7 mm in diameter), macronodules (> 7 mm in diameter), silicotic nodules, progressive massive fibrosis (PMF), Caplan's lesions, and infective granulomas (e.g., in histoplasmosis and tuberculosis). The nodular lesions are palpable and gray to black, and they contain collagen, silica, and silicates, as well as black particles. The silicotic nodules are composed of whorled or laminated collagen and silica and can also contain some black particles. The lesions in PMF are black and rubbery to hard, and they can contain inky fluid. One of the criteria for defining PMF lesions is a minimal size, 1–3 cm (Kleinerman et al., 1979).

One of the major subjects of controversy in CWP is whether coal-mine dust causes emphysema and, if it does, whether it is clinically significant. In early reports, two British pathologists considered focal emphysema around the coal macules to "constitute the characteristic feature of the disease" (Gough, 1947) or to represent "the most important feature of the larger lesions," macules 2–4 mm in diameter (Heppleston, 1947). The researchers did not try to relate the extent of the emphysema to clinical signs and symptoms. Heppleston (1947) did note, however, that "focal emphysema is occasionally so extensive as to leave but little of the parenchyma unaffected."

It was not until recently that controlled studies were performed in an attempt to determine whether coal miners exhibited more emphysema than other workers. Cockcroft et al. (1982), in an autopsy series on 39 coal workers and 48 other workers, found that coal workers who smoked exhibited much more centrilobular emphysema than controls who smoked. There were too few nonsmoking coal workers in the sample to determine whether work in coal mines by itself could produce significant emphysema. Ruckley et al. (1984) were able to show, in a series of 450 miners, that centriacinar emphysema was significantly more prevalent in smoking (72%) than in nonsmoking (42%) miners. They also showed that increasing concentrations of lung dust correlated with a higher incidence of centriacinar emphysema ($p < 0.05$). Although there were no nonsmoking, nonminer controls, the fact that a high percentage (75%) of the nonsmoking miners with PMF had emphysema strongly suggests that nonsmoking miners with PMF have a much higher risk of developing emphysema than do nonsmoking nonminers. Ryder et al. (1970), in a study of 247 coal miners, were able to show a highly significant correlation between extent of emphysema and a decrease in FEV_1. It seems clear that coal miners have a higher risk of emphysema than do other workers; however, this trend has been clearly demonstrated only for smokers and possibly for nonsmoking coal miners with PMF.

A small but statistically significant decrease in FEV_1 in active coal miners has also been demonstrated. Rogan et al. (1973) reported a statistically significant ($p < 0.001$) progressive reduction in FEV_1 with increasing cumulative exposure to coal-mine dust in a prospective study on 3,581 coal-face workers. In men without PMF, the reduction in FEV_1 was calculated to be 120–150 ml over 35 years at mean respirable-dust concentrations of 4 mg/m^3. The authors equated that with the loss one might expect from smoking 20 cigarettes/day for the same period. The decrements in FEV_1 were seen in nonsmokers, as well as in smokers. They found no interaction (synergism) between cigarette-smoking and dust exposure. However, the greatest decrement in FEV_1 was found in workers with chronic bronchitis, and it could not be explained solely on the basis of dust exposure and cigarette-smoking. In a longitudinal study of 1,677 active coal miners without PMF, Love and Miller (1982) found that the loss of FEV_1 over approximately 11 years increased with cumulative dust exposure, when the effects of age, height, and smoking were allowed for. They estimated that exposure to dust at the average concentration (117 $g \times hours/m^3$) recorded for the 1,677 men studied was associated with about a 40-ml loss in FEV_1 over an 11-year period. Lloyd (1971) was able to show a significant difference in pulmonary function between active miners and a nonmining control group. Of current miners, 20% had an FEV_1 less than 80% of predicted, compared with 10% of active telecommunication workers.

Hankinson et al. (1977) demonstrated a statistically significant correlation between decrements in flow rates at high lung volumes and years of underground exposure. The decrement was more noticeable among the nonsmokers. They suggested that a dust-induced bronchitis was responsible. Those studies, as well as others, have shown that the pulmonary function deficits are rather small (except in workers with PMF). Most of the studies focused on active miners and therefore might have analyzed a self-selected group of people who can tolerate dust exposures without significant effects on pulmonary function. Some support for that idea was recently published by Hurley and Soutar (1986). In a study of 199 men without PMF who had left the coal industry before normal retirement age, a much greater loss of FEV_1 than expected was found (average loss, 600 ml in those who had experienced moderately high exposures).

A higher incidence of chronic bronchitis also has been shown to occur in coal miners than in nonminers (Lloyd, 1971). Because the sample contained so few nonsmokers, however, the comparison was statistically significant only in the smoking group. Rae et al. (1971), in a study of 4,122 face workers from 20 collieries, found a statistically significant association between increasing exposure to coal dust and increasing prevalence of bronchitis in the 25–34 and 35–44 age groups. That was true in both nonsmokers and smokers. Coal-mine dust exposure was also found to be significantly related to maximal ratio of mucous-gland thickness to bronchial-wall thickness, an anatomic index of chronic bronchitis (Douglas et al., 1982).

The most common method for evaluating CWP in the living is with diagnostic chest x rays. A rather elaborate system has been adopted internationally for the grading of x-ray changes in coal workers. CWP is divided into simple CWP and complicated CWP, or PMF. Simple CWP includes all cases in which the x-ray opacities (rounded or irregular) are less than 1 cm in diameter and is subdivided into several categories that depend on the abundance ("profusion") of these opacities. Jacobsen et al. (1971), in a prospective study of 4,122 coal workers over a 10-year period, found a linear correlation between the progression of chest x-ray category and exposure, measured as the concentration of respirable coal-mine dust per cubic meter. They estimated that exposures to coal-mine dust at 2 mg/m^3 for 35 years would lead to fewer than 1% of miners progressing from category 0/0 (no CWP) to category 2 or higher.

On the basis of the studies noted here, it is clear that coal miners exhibit a higher incidence of bronchitis, pulmonary function deficits, and emphysema than nonminers. It is not clear to what extent the changes are caused by coal-mine dust, cigarette-smoking, a combination of coal-mine dust and cigarette-smoking, or other factors.

The composition of coal dust and coal-mine dust might vary considerably, depending on the type of coal and the geologic area of the coal seams. Typical constituents in coal in the United States (Schlick and Fannick, 1971) are shown carbon (29.5–81%), volatile matter (5.1–36.8%), moisture (4.3–36.8%), ash (4.3–9.6%), and sulfur (0.5–0.9%). The ash can contain quartz, aluminum silicates (e.g., mica and kaolin), and iron oxides in substantial concentrations. Coal-mine dusts might contain a variety of other minerals, depending on how much and what kind of rock dust is generated in the process of roof-bolting and cutting the coal seams. In addition, miners are occasionally exposed to fumes from cable fires and explosives.

Faced with this wide range of variables and limited resources, most investigators studied what they considered to be the most active ingredient in coal-mine dust—quartz. The pulmonary effects of coal dusts that have various concentrations of mineral matter and of quartz have been studied sporadically in animal systems for more than 50 years. The principal end points in the investigations were the development of macules and fibrotic nodules in the lungs of the experimental animals.

The results of inhalation studies in animals have generally shown that coal dust containing less than 10% quartz has not produced substantial pulmonary fibrosis. Ross et al. (1962) exposed rats to a mixed dust of anthracite coal with a low ash content and quartz added in various concentrations. Exposure was to respirable dust at about 60 mg/m^3 for about 3,200 hours over 10–17 months. Rats exposed to mixtures of coal dust with 5% and 10% quartz showed little or no pulmonary fibrosis, whereas those exposed to coal dust with 20% quartz showed grade 2 fibrosis, and those exposed to coal dust with 40% quartz showed grade 4 fibrosis (maximal grade, 5). Weller and Ulmer (1972) exposed animals to coal dust at 45 mg/m^3 containing 30–35% added quartz for 1,800–2,200 hours over 18 months (rats) and 3 years (monkeys). The rats developed grade 2 fibrosis, and the monkeys developed grade 2–3 fibrosis. The monkeys developed macules that looked similar to those seen in coal miners' lungs. Rats did not develop macules with the same appearance as those in coal miners, probably because the anatomy of the rat lung is considerably different from that of the human lung. Gross and Nau (1967) exposed monkeys, rats, guinea pigs, and mice to lignite coal dust at 7.8 mg/m^3 containing 2% quartz and carbon dust at 8.1 mg/m^3 containing 6.5% quartz for 1,820 hours over 1 year. They observed no fibrosis and little or no reticulin in the dust aggregates in either experiment. They observed that the lungs of mice contained less dust than the lungs of rats, which contained less dust than the lungs of monkeys. Martin et al. (1977) exposed rats to coal dust at 200 mg/m^3 or with 10% quartz for up to 1,300 hours over 2 years. They observed small amounts of reticulin (but no fibrosis) in the dust masses in the lungs of rats exposed to coal dust and massive nodular lesions with collagen fibers in the lungs of rats exposed to coal dust that contained 10% quartz. The nodules looked similar to fibrotic nodules seen in the lungs of coal miners with CWP, but did not look like the classical silicotic nodules composed of laminated or whorled collagen.

The effects of the minerals in coal dust on the fibrogenicity of quartz have also been studied. In relatively short-term studies, Martin et al. (1972) have shown that mineral matter from coal dust can partially inhibit the fibrogenicity of quartz. They also found that quartz treated with a dialysate from coal mineral matter has a significantly reduced solubility and induced much less collagen in the lungs of rats in 3-month experiments. That raises the possibility that quartz in different coal-mine dusts might have different capacities to induce fibrotic changes in the lungs of miners and that the capacity might vary with the composition of the mineral matter that is inhaled with it. Lungs of coal miners might contain quartz at relatively high concentrations (Davis et al., 1983). Lungs with macules but no fibrotic nodules contained quartz at a mean concentration of 800 mg per lung—the same mean concentration seen in the lungs of workers with slight silicosis (Nagelschmidt, 1965.) The lungs also contained 3,300 mg of kaolin and mica, however. The high concentration of those minerals might have modified the fibrogenic effect of the quartz.

Few pulmonary function studies in animals have been reported. Moorman et al. (1977) exposed germ-free and conventional rats to coal dust at 10 mg/m^3 for 960 hours. They showed small but statistically significant reductions in flow maximums at 10% of vital capacity in the exposed rats, compared with the controls. Weller and Ulmer (1972) performed pulmonary function tests on rats exposed to coal dust at 45 mg/m^3 (with 30–35% quartz) for 18 months. They found small but statistically significant differences only in the values for compliance.

No experimental models have been developed to study chronic bronchitis and centrilobular emphysema as they occur in humans.

REFERENCES

Alavanja, M., I. Goldstein, and M. Susser. 1978. A case control study of gastrointestinal and urinary tract cancer mortality and drinking water chlorination, pp. 395–409. In R. L. Jolley, H. Gorchev, and D. H. Hamilton, Jr. (eds.). Water Chlorination: Environmental Impact and Health Effects. Vol. 2. Ann Arbor Science, Ann Arbor, Mich.

Amdur, M. O. 1985. When one plus zero is more than one. Am. Ind. Hyg Assoc. J. 46:467–475.

Amdur, M. O., and J. Mead. 1958. Mechanics of respiration in unanesthetized guinea pigs. Am. J. Physiol. 192:364–368.

Amdur, M. O., and D. Underhill. 1968. The effect of various aerosols on the response of guinea pigs to sulfur dioxide. Arch. Environ. Health 16:460–468.

Archer, V. E. 1985. Enhancement of lung cancer by cigarette smoking in uranium and other miners, pp. 23–37. In M. J. Mass (ed.). Carcinogenesis. Vol. 8. Raven Press, New York.

Axelson, O., and L. Sundell. 1978. Mining, lung cancer and smoking. Scand. J. Work Environ. Health 4:46–52.

Balster, R. L., and J. F. Borzelleca. 1982. Behavioral toxicity of trihalomethane contaminants of drinking water in mice. Environ. Health Perspect. 46:127–136.

Batuman, V., E. Landy, J. K. Maesaka, and R. P Wedeen. 1983. Contribution of lead to hypertension with renal impairment. N. Engl. J. Med. 309:17–21.

Begin, R., A. Cantin, Y. Berthiaume, R. Boileau, S. Peloquin, and S. Masse. 1983. Airway function in lifetime-nonsmoking older asbestos workers. Am. J. Med. 75:631–638.

Bellar, T. A., J. J. Lichtenberg, and R. C. Kroner. 1974. The occurrence of organohalides in chlorinated drinking waters. J. Am. Water Works Assoc. 66:703–706.

Berry, G., M. L. Newhouse, and M. Turok. 1972. Combined effect of asbestos exposure and smoking on mortality from lung cancer in factory workers. Lancet 2:476–478.

Brenniman, G. R., J. Vasilomanolakis-Lagos, J. Amsel, T. Namekata, and A. H. Wolff. 1980. Case-control study of cancer deaths in Illinois communities served by chlorinated or nonchlorinated water, pp. 1043–1057. In R. L. Jolley, W. A. Brungs, R. B. Cumming, and V. A. Jacobs (eds.). Water Chlorination: Environmental Impact and Health Effects. Vol. 3. Ann Arbor Science, Ann Arbor, Mich.

Buechley, R. W., W. B. Riggan, V. Hasselblad, and J. B. VanBruggen. 1973. SO_2 levels and perturbations in mortality: A study in the New York-New Jersey metropolis. Arch. Environ. Health 27: 134–137.

Burke, T. A., J. Amsel, and K. P. Cantor. 1983. Trihalomethane variation in public drinking water supplies, pp. 1343–1351. In R. L. Jolley, W. A. Brungs, J. A. Cotruvo, R. B. Cumming, J. S. Mattice, and V. A. Jacobs (eds.). Water Chlorination: Environmental Impact and Health Effects. Vol. 4, Book 2: Environment, Health, and Risk. Ann Arbor Science, Ann Arbor, Mich.

Cantor, K. P., R. Hoover, P. Hartge, T. J. Mason, D. T. Silverman, and L. I. Levin. 1985. Drinking water source and risk of bladder cancer: A case-control study, pp. 145–152. In R. L. Jolley, R. J. Bull, W. P. Davis, S. Katz, M. H. Roberts, Jr., and V. A. Jacobs (eds.). Water Chlorination: Chemistry, Environmental Impact and Health Effects. Vol. 5. Lewis Publishers, Chelsea, Mich.

Cockcroft, A., R. M. E. Seal, J. C. Wagner, J. P. Lyons, R. Ryder, and N. Andersson. 1982. Post-mortem study of emphysema in coalworkers and non-coalworkers. Lancet 2(8298):600–603.

Cragle, D. L., C. M. Shy, R. J. Struba, and E. J. Siff. 1985. A case-control study of colon cancer and water chlorination in North Carolina, pp. 153–159. In R. L. Jolley, R. J. Bull, W. P. Davis, S. Katz, M. H. Roberts, Jr., and V. A. Jacobs (eds.). Water Chlorination: Chemistry, Environmental Impact and Health Effects. Vol. 5. Lewis Publishers, Chelsea, Mich.

Craun, G. F. 1985. Epidemiologic considerations for evaluating associations between the disinfection of drinking water and cancer in humans, pp. 133–143. In R. L. Jolley, R. J. Bull, W. P. Davis, S. Katz, M. H. Roberts, Jr., and V. A. Jacobs (eds.). Water Chlorination: Chemistry, Environmental Impact and Health Effects. Vol. 5. Lewis Publishers, Chelsea, Mich.

Cross, F. T., R. F. Palmer, R. E. Filipy, G. E. Dagle, and B. O. Stuart. 1982. Carcinogenic effects of radon daughters, uranium ore dust and cigarette smoke in beagle dogs. Health Phys. 42:33–52.

Damber, L. and L. G. Larsson. 1985. Underground mining, smoking, and lung cancer: A case-control study in the iron ore municipalities in northern Sweden. J. Natl. Cancer Inst. 74:1207–1213.

Davis, J. M. G., S. T. Beckett, R. E. Bolton, and K. Donaldson. 1980. A comparison of the pathological effects in rats of the UICC reference samples of amosite and chrysotile with those of amosite and chrysotile collected from the factory environment, pp. 285–292. In J. C. Wagner (ed.). Biological Effects of Mineral Fibers. Vol. 1. IARC Scientific Pub. No. 30. International Agency for Research on Cancer, Lyon, France.

Davis, J. M. G., J. Chapman, P. Collings, A. N. Douglas, J. Fernie, D. Lamb, and V. A. Ruckley. 1983. Variations in the histological patterns of the lesions of coal workers' pneumoconiosis in Britain and their relationship to lung dust content. Am. Rev. Respir. Dis. 128:118–124.

DeRouen, T. A., and J. E. Diem. 1977. Relationships between cancer mortality in Louisiana drinking-water source and other possible causative agents, pp. 331–356. In H. H. Hiatt, J. D. Watson, and

J. A. Winsten (eds.). Origins of Human Cancer. Book A: Incidence of Cancer in Humans. Cold Spring Harbor Laboratory, Cold Spring Harbor, N.Y.

DeShazo, R. D., D. J. Hendrick, J. E. Diem, J. A. Nordberg, Y. Baser, D. Brevier, R. N. Jones, H. W. Barkman, J. E. Salvaggio, and H. Weill. 1983. Immunologic aberrations in asbestos cement workers: Dissociation from asbestosis. J. Allergy Clin. Immunol. 72:454–461.

Doll, R. 1955. Mortality from lung cancer in asbestos workers. Br. J. Ind. Med. 12:81–86.

Douglas, A. N., D. Lamb, and V. A. Ruckley. 1982. Bronchial gland dimensions in coalminers: Influence of smoking and dust exposure. Thorax 37:760–764.

Edling, C. 1982. Lung cancer and smoking in a group of iron ore miners. Am. J. Ind. Med. 3:191–199.

Gottlieb, M. S., and J. K. Carr. 1982. Case-control cancer mortality study and chlorination of drinking water in Louisiana. Environ. Health Perspect. 46:169–177.

Gough, J. 1947. Pneumoconiosis in coal workers in Wales. Occup. Med. 4:86–97.

Great Britain, Committee on Air Pollution. 1954. Report. Cmd 9322. Her Majesty's Stationery Office, London.

Gross, P. and C. A. Nau. 1967. Lignite and the derived steam-activated carbon: The pulmonary response to their dusts. Arch. Environ. Health 14:450–460.

Hammond, E. C., I. J. Selikoff, and H. Seidman. 1979. Asbestos exposure, cigarette smoking and death rates. Ann. N.Y. Acad. Sci. 330:473–490.

Hankinson, J. L., R. B. Reger. R. P. Fairman, N. L. Lapp, and W. K. C. Morgan. 1977. Factors influencing expiratory flow rates in coal miners, pp. 737–755. In W. H. Walton (ed.). Inhaled Particles IV. Part 2. Pergamon, New York.

Harlan, W. R., J. R. Landis, R. L. Schmouder, N. G. Goldstein, and L. C. Harlan. 1985. Blood lead and blood pressure: Relationship in the adolescent and adult U.S. population. J. Am. Med. Assoc. 253:530–534.

Harries, R. G., F. A. F. MacKenzie, G. Sheers, J. H. Kemp, T. P. Oliver, and D. S. Wright. 1972. Radiological survey of men exposed to asbestos in naval dockyards. Br. J. Ind. Med. 29:274–279.

Heppleston, A. G. 1947. The essential lesion of pneumoconiosis in Welsh coal workers. J. Pathol. Bacteriol. 59:453–460.

Hornung, R. W., and S. Samuels. 1981. Survivorship models for lung cancer mortality in uranium miners: Is cumulative dose an appropriate measure of exposure?, pp. 363–368. In M. Gomez (ed.). International Conference—Radiation Hazards in Mining: Control, Measurement, and Medical Aspects, Golden, Colo., October 4-9, 1981. Society of Mining Engineers of the AIME, New York.

Hurley, J. F., and C. A. Soutar. 1986. Can exposure to coalmine dust cause a severe impairment of lung function? Br. J. Ind. Med. 43:150–157.

International Agency for Research on Cancer (IARC). 1985. IARC Monographs on the Evaluation of the Carcinogenic Risk of Chemicals to Humans. Volume 35: Polynuclear Aromatic Compounds. Part 4: Bitumens, Coal-Tars and Derived Products, Shale-Oils and Soots. International Agency for Research on Cancer, Lyon, France.

Jacobsen, M., S. Rae, W. H. Walton, and J. M. Rogan. 1971. The relation between pneumoconiosis and dust-exposure in British coal mines, pp. 903–919. In W. H. Walton (ed.). Inhaled Particles III, Vol. II. Unwin Brothers, Surrey, England.

Johnson, G. A. 1911. Hypochlorite treatment of public water supplies: Its adaptability and limitations. J. Am. Public Health Assoc. 1:562–574.

Kanarek, M. S., and T. B. Young. 1982. Drinking water treatment and risk of cancer death in Wisconsin. Environ. Health Perspect. 46:179–186.

Keller A. Z., and M. Terris. 1965. The association of alcohol and tobacco with cancer of the mouth and pharynx. Am. J. Public Health 55:1578–1585.

Kennaway, N. M., and E. L. Kennaway. 1936. Study of the incidence of cancer of the lung and larynx. J. Hyg. 36:236–267.

Khera, A. K., D. G. Wibberley, K. W. Edwards, and H. A. Waldron. 1980. Cadmium and lead levels in blood and urine in a series of cardiovascular and normotensive patients. Int. J. Environ. Stud. 14:309–312.

Kleinerman, J., F. Green, R. A. Harley, N. L. Lapp, W. Laqueur, R. Naeye, P. Pratt, G. Taylor, J. Wiot, and J. Wyatt. 1979. Pathology standards for coal workers' pneumoconiosis. Arch. Pathol. Lab. Med. 103:375–432.

Kromhout, D., and C. de Lezenne Coulander. 1984. Trace metals and CHD risk indicators in 152 elderly men (the Zutphen study). Eur. Heart J. 5 (Abstr. Suppl. 1):101.

Kromhout, D., A. A. E. Wibowo, R. F. M. Herber, L. M. Dalderup, H. Heerdink, C. de Lezenne Coulander, and R. L. Zielhuis. 1985. Trace metals and coronary heart disease risk indicators in 152 elderly men (the Zutphen study). Am. J. Epidemiol. 122:378–385.

Kuzma, R. J., C. M. Kuzma, and C. R. Buncher. 1977. Ohio drinking water source and cancer rates. Am. J. Public Health 67:725–729.

Lange, A. 1980. An epidemiological survey of immunological abnormalities in asbestos workers. II. Serum immunoglobulin levels. Environ. Res. 176–183.

Liddell, F. D. K., G. W. Gibbs, and J. C. McDonald. 1982. Radiological changes and fibre exposure in chrysotile workers aged 60–69 years at Thetford Mines. Ann. Occup. Hyg. 26:889–898.

Lippmann, M. 1985. Airborne acidity: Estimates of exposure and human health effects. Environ. Health Perspect. 63:63–70.

Lippmann, M., R. B. Schlesinger, G. Leikauf, D. Spektor, and R. E. Albert. 1982. Effects of sulphuric acid aerosols on respiratory tract airways. Ann. Occup. Hyg. 26:677–690.

Lloyd, I. W. 1971. Long-term mortality study of steelworkers. V. Respiratory cancer in coke plant workers. J. Occup. Med. 13:53–68.

Love, R. G., and B. G. Miller. 1982. Longitudinal study of lung function in coal-miners. Thorax 37:193–197.

Lynch, K. M., and W. A. Smith. 1935. Pulmonary asbestosis III: Carcinoma of lung in asbesto-silicosis. Am. J. Cancer. 24:56–64.

Martin, J. C., H. Daniel-Moussard, L. Le Bouffant, and A. Policard. 1972. The role of quartz in the development of coal workers' pneumoconiosis. Ann. N.Y. Acad. Sci. 200:127–141.

Martin, J. C., H. Daniel, and L. Le Bouffant. 1977. Short- and long-term experimental study of the toxicity of coal-mine dust and some of its constituents, pp. 361–371. In W. H. Walton (ed.). Inhaled Particles IV. Part 1. Pergamon, New York.

McDonald, J. C., F. D. K. Liddell, G. W. Gibbs, G. E. Eyssen, and A. D. McDonald. 1980. Dust exposure and mortality in chrysotile mining, 1910–75. Br. J. Ind. Med. 37:11–24.

McGregor, J. F. 1976. Tumor-promoting activity of cigarette tar in rat skin exposed to irradiation. J. Natl. Cancer Inst. 56:429–430.

McJilton, C. E., R. Frank, and R. J. Charlson. 1976. Influence of relative humidity on functional effects of an inhaled SO_2-aerosol mixture. Am. Rev. Respir. Dis. 113:163–169.

Meier, J. R., H. P. Ringhand, W. E. Coleman, J. W. Munch, R. P. Streicher, W. H. Kaylor, and K. M. Schenck. 1985. Identification of mutagenic compounds formed during chlorination of humic acid. Mutat. Res. 157:111–122.

Meurman, L. O., R. Kiviluoto, and M. Hamaka. 1974. Mortality and morbidity among the working population of anthophyllite asbestos miners in Finland. Br. J. Ind. Med. 31:105–112.

Moorman, W. J., R. W. Hornung, and W. D. Wagner. 1977. Ventilatory functions in germfree and conventional rats exposed to coal dusts. Proc. Soc. Exp. Biol. Med. 155:424–428.

Moreau, T., G. Orssaud, B. Juguet, and G. Busquet. 1982. Blood lead levels and arterial pressure: Initial results of a cross sectional study of 431 male subjects. Rev. Epidemol. Sante Publique 30:395–397. (In French.)

Mossman, B. T., and J. E. Craighead. 1982. Comparative cocarcinogenic effects of crocidolite asbestos, hematite, kaolin and carbon in implanted tracheal organ cultures. Ann. Occup. Hyg. 26: 553–567.

Nagelschmidt, G. 1965. The study of lung dust in pneumoconiosis. Am. Ind. Hyg. Assoc. J. 26:1–7.

Nenot, J. C. 1977. In IAEA. Proceedings of the IAEA Symposium. International Atomic Energy Agency, Chicago.,

Newhouse, M. L., G. Berry, J. C. Wagner, and M. E. Turok. 1972. A study of the mortality of female asbestos workers. Br. J. Ind. Med. 29:134–141.

Ozkaynak, H., and J. D. Spengler. 1985. Analysis of health effects resulting from population exposures to acid precipitation precursors. Environ. Health Perspect. 63:45–55.

Page, N. P., and U. Saffiotti. 1976. Report on Carcinogenesis Bioassay of Chloroform. National Cancer Institute, Division of Cancer Cause and Prevention, Bethesda, Md. (60 pp.)

Pirkle, J. L., J. Schwartz, J. R. Landis, and W. R. Harlan. 1985. The relationship between blood lead levels and blood pressure and its cardiovascular risk implications. Am. J. Epidemiol. 121:246–258.

Pocock, S. J., A. G. Shaper, D. Ashby, T. Delves, and T. P. Whitehead. 1984. Blood lead concentration, blood pressure, and renal function. Br. Med. J. 289:872–874.

Pocock, S. J., A. G. Shaper, D. Ashby, and T. Delves. 1985. Blood lead and blood pressure in middle-aged men, pp. 303–305. In T. D. Lekkas (ed.). International Conference: Heavy Metals in the Environment. Vol. 1. September; Athens, Greece. CEP Consultants, Edinburgh, United Kingdom.

Pott, P. 1775. Cancer scroti, pp. 63–68. In Chirurgical Observations Relative to the Cataract, the Polypus of the Nose, the Cancer of the Scrotum, the Different Kinds of Ruptures, and the Mortification of the Toes and Feet. Printed by T. J. Carnegy for L. Hawes, W. Clarke, and R. Collins, London.

Radford, E. P., and R. G. St. Clair Renard. 1984. Lung cancer in Swedish iron miners exposed to low doses of radon daughters. New Engl. J. Med. 310:1485–1494.

Rae, S., D. D. Walker, and M. D. Attfield. 1971. Chronic bronchitis and dust exposure in British coalminers, pp. 883–896. In W. H. Walton (ed.). Inhaled Particles III. Volume II. Unwin Brothers, Surrey, England.

Regan, G. M., B. Tagg. J. Walford, and M. L. Thomson. 1971. The relative importance of clinical, radiological and pulmonary function variables in evaluating asbestosis and chronic obstructive airway disease in asbestos workers. Clin. Sci. 41:569–582.

Rogan, J. M., M. D. Attfield, M. Jacobsen, S. Rae, D. D. Walker, and W. H. Walton. 1973. Role of dust in the working environment in development of chronic bronchitis in British coal miners. Br. J. Ind. Med. 30:217–226.

Rook, J. J. 1974. Formation of haloforms during chlorination of natural waters. Water Treat. Exam. 23:234–243.

Ross, H. F., E. J. King, M. Yoganathan, and G. Nagelschmidt. 1962. Inhalation experiments with coal dust containing 5 per cent, 10 per cent, 20 per cent and 40 per cent quartz: Tissue reactions in the lungs of rats. Ann. Occup. Hyg. 5:149–161.

Rothman K., and A. Keller. 1972. The effect of joint exposure to alcohol and tobacco on risk of cancer of the mouth and pharynx. J. Chron. Dis. 25:711–716.

Ruckley, V. A., S. J. Gauld, J. S. Chapman, J. M. G. Davis, A. N. Douglas, J. M. Fernie, M. Jacobsen, and D. Lamb. 1984. Emphysema and dust exposure in a group of coal workers. Am. Rev. Respir. Dis. 129:528–532.

Ryder, R., Lynons, J. P., H. Campbell, and J. Gough,. 1970. Emphysema in coal workers' pneumoconiosis. Br. Med. J. 3:481–487.

Salg, J. 1977. Cancer Mortality Rates and Drinking Water Quality in the Ohio River Valley Basin. Ph.D. dissertation, University of North Carolina at Chapel Hill. (144 pp.)

Saracci, R. 1977. Asbestos and lung cancer: An analysis of the epidemiological evidence on the asbestos-smoking interaction. Int. J. Cancer 20:323–331.

Schimmel, H. 1978. Evidence for possible acute health effects of ambient air pollution from time series analysis: Methodological questions and some new results based on New York City daily mortality, 1963–1976. Bull. N.Y. Acad. Med. 54:1052–1108.

Schimmel, H., and T. J. Murawski. 1975. SO_2—Harmful pollutant or air quality indicator? J. Air Pollut. Control Assoc. 25:739–740.

Schimmel, H., and T. J. Murawski. 1976. The relation of air pollution to mortality. J. Occup. Med. 18:316–333.

Schlesinger, R. B. 1985. The effects of inhaled acids on respiratory tract defense mechanisms. Environ. Health Perspect. 63:25–38.

Schlick, D. P., and N. L. Fannick. 1971. Coal in the United States, pp. 13–26. In M. M. Key, L. E. Kerr, and M. Bundy (eds.). Pulmonary Reactions to Coal Dust: A Review of U.S. Experience. Academic Press, New York.

Seidman, H., I. J. Selikoff, and E. C. Hammond. 1979. Short-term asbestos work exposure and long-term observation. Ann. N.Y. Acad. Sci. 330:61–89.

Selikoff, I. J., E. C. Hammond, and J. Churg. 1968. Asbestos exposure, smoking, and neoplasia. J. Am. Med. Assoc. 204:106–112.

Sheppard, D., W. S. Wong, C. F. Uehara, J. A. Nadel, and H. A. Boushey. 1980. Lower threshold and greater bronchomotor responsiveness of asthmatic subjects to sulfur dioxide. Am. Rev. Respir. Dis. 122:873–878.

Stevens, A. A., R. C. Dressman, R. K. Sorrell, and H. J. Brass. 1985. Organic halogen measurements: Current uses and future prospects. J. Am. Water Works Assoc. 77(4):146–154.

Topping, D. C., and P. Nettesheim. 1980. Two-stage carcinogenesis studies with asbestos in Fischer 344 rats. J. Natl. Cancer Inst. 65:627–630.

U.S. EPA (Environmental Protection Agency), Environmental Criteria and Assessment Office. 1980. Long-Term Effects of Air Pollutants in Canine Species. EPA-600/8-80-014. U.S. EPA, Cincinnati, Ohio.

U.S. EPA, Environmental Criteria and Assessment Office. 1982. Air Quality Criteria for Particulate Matter and Sulfur Oxides. Vol. III. EPA-600/8-82-029a. U.S. EPA, Research Triangle Park, N.C.

U.S. EPA, Environmental Criteria and Assessment Office. 1984. Air Quality Criteria for Lead, Review Draft. EPA-600/8-83-028. U.S. EPA, Research Triangle Park, N.C. (4 vols.)

U. S. Public Health Service, Division of Water Supply and Pollution Control. 1964. Inventory of Municipal Water Facilities (1963). Region V. Illinois, Indiana, Michigan, Ohio, and Wisconsin. PHS Publication No. 775. Vol. 5. U. S. Government Printing Office, Washington, D.C. (Available from NTIS as PB-218194.) (152 pp.)

Utell, M. J. 1985. Effects of inhaled acid aerosols on lung mechanics: An analysis of human exposure studies. Environ. Health Perspect. 63:39–44.

Victery, W., A. J. Vander, H. Markel, L. Katzman, J. M. Shulak, C. Germain. 1982a. Lead exposure, begun *in utero*, decreases renin and angiotensin II in adult rats (41398). Proc. Soc. Exp. Biol. Med. 170:63–67.

Victery, W., A. J. Vander, J. M. Shulak, P. Schoeps, and S. Julius. 1982b. Lead, hypertension, and the renin-angiotensin system in rats. J. Lab. Clin. Med. 99:354–362.

Wagner, M. M. F. 1980. Immunology and asbestos, pp. 247–251. In J. C. Wagner (ed.). Biological Effects of Mineral Fibers. Vol. 2. IARC Scientific Pub. No. 30. International Agency for Research on Cancer, Lyon, France.

Wagner, J. C., C. A. Sleggs, and P. Marchand. 1960. Diffuse pleural mesothelioma and asbestos exposure in the North Western Cape Province. Br. J. Ind. Med. 17:260–271.

Wagner, J. C., G. Berry, J. W. Skidmore, and F. D. Pooley. 1980. The comparative effects of three chrysotiles by injection and inhalation in rats, pp. 363–372. In J. C. Wagner (ed.). Biological Effects of Mineral Fibers. Vol. 1. IARC Scientific Pub. No. 30. International Agency for Research on Cancer, Lyon, France.

Walker, A. M. 1984. Declining relative risks for lung cancer after cessation of asbestos exposure. J. Occup. Med. 26:422–426.

Wehner, A. P., R. H. Busch, R. J. Olson, and D. K. Craig. 1975. Chronic inhalation of asbestos and cigarette smoke by hamsters. Environ. Res. 10:368–383.

Weiss, S. T., A. Munoz, A. Stein, D. Sparrow, and F. E. Speizer. 1986. The relationship of blood lead to blood pressure in a longitudinal study of working men. Am. J. Epidemiol. 123:800–808.

Weller, W., and W. T. Ulmer. 1972. Inhalation studies of coal-quartz dust mixture. Ann. N.Y. Acad. Sci. 200:142–154.

Wright, J. L., and A. Churg. 1984. Morphology of small-airway lesions in patients with asbestos exposure. Hum. Pathol. 15:68–74.

Wynder E. L., I. J. Bross, and R. M. Feldman. 1957. A study of the etiological factors in cancer of the mouth. Cancer 10:1300–1323.

Yamagiwa, K., and K. Ichikawa. 1918. Experimental study of the pathogenesis of carcinoma. J. Cancer Res. 3:1–29.

Young, T. B., and M. S. Kanarek. 1983. Matched pair case control study of drinking water chlorination and cancer mortality, pp. 1365–1380. In R. L. Jolley, W. A. Brungs, J. A. Cotruvo, R. B. Cumming, J. S. Mattice, and V. A. Jacobs (eds.). Water Chlorination: Environmental Impact and Health Effects. Vol. 4, Book 2: Environment, Health, and Risk. Ann Arbor Science, Ann Arbor, Mich.

C

Case Studies on Strategies for Testing the Toxicity of Complex Mixtures

We present here several case studies that involve testing of the toxicity of complex mixtures. A data base on each of these mixtures existed before any formalized system of testing or strategies, and it is instructive to look at the cases with the aid of the strategies presented in Chapter 3. How was the problem defined? Were the data developed along predetermined lines? Was a strategy evident? Was it substantially different from that proposed in this report? Was it successful? Might another strategy have been of more value?

The cases chosen represent a range of mixtures and a range of toxic end points. The end points were carcinogenesis and mutagenesis caused by cigarette smoke and fractions, systemic toxicity of fire atmospheres, and neuropathy associated with hydrocarbon and oxygenated solvents.

The reader is encouraged to review these cases in the light of the strategies used—that is, with respect to their success and to whether any surprises resulted from asking the wrong questions (as in the case of species specificity of sedation by thalidomide) or from failure to ask any questions at all (as in the case of unplanned release of methyl isocyanate from bulk storage).

One further purpose of these case studies is to show that complex mixtures can and should be addressed by looking at basic and simple elements. Rather than being concerned with finding the perfect approach to a problem, we should look for feasible approaches. Complex mixtures can be thought of as representing laboratory questions in terms of composition and exposure.

CIGARETTE-SMOKE TOXICITY

Careful observation of the effects of cigarette-smoke inhalation in humans teaches many lessons. Among the most striking observations are the following:

- The large number of agents in cigarette smoke (more than 3,600).
- The development of both chronic and acute effects—most unusual is the unpredictable and diverse character of the chronic lesions, which cannot be foretold from the nature of the acute injury.
- The variety of organ systems that appear to be affected.
- The synergistic interactions that can occur between cigarette smoke and environmental or occupational agents and accelerate or alter pathologic change.

More than 3,600 components are generated in the distillation zone distal to the combustion zone. Mainstream smoke is drawn into the smoker's mouth from the burning cigarette; sidestream smoke arises from the burning end and is released into the environment. The composition of mainstream and sidestream smoke varies with the smoking conditions.

Standard cigarette-smoking conditions are created to produce an average for a male smoker of a nonfiltered cigarette. For filtered cigarettes, the average puff lasts 1.94–2.06 seconds and is repeated every 26.9–30.0 seconds, and the puff volume is 35.9–47.8 ml. For cigar-smoking, the standard conditions used are a 1.5-second puff every 40 seconds with a volume of 20 ml and a butt length of 33 mm. For pipe-smokers, the standardization is less rigorous, but 2-second puffs every 18 seconds with a volume of 50 ml have often been used. Only 30% of the mainstream-smoke weight originates in the tobacco; the remaining weight is from the air drawn in with the smoke. Of the effluent of a nonfiltered cigarette, 5–9% by weight is moist particulate matter, 55–65% is nitrogen gas, 8–14% is oxygen, and the remainder is other gas-phase material. Undiluted cigarette smoke at the mouth contains 5×10^9 particles/ml in a size range of 0.2–1.0 mm (median particle size, 0.4 μm). For filtered cigarettes, the particle size is 0.35–0.4 μm; for cigarettes with perforated filter tips, the particle number is substantially lower. The smoke particles are charged with 10^{12} electrons per gram of smoke and have reducing activity when freshly inhaled.

The gaseous phase of tobacco smoke does not induce malignant tumors of the respiratory tract in laboratory animals. That observation suggests that the carcinogenic activity of smoke requires the particulate phase. Benign and malignant tumors have been induced with tobacco tar in skin and ear of rabbits, by intratracheal instillation in rats, and by topical application on mouse skin. Experimental studies with classically defined initiators and promoters have demonstrated that the effect of a tumor initiator is irreversible, but promoters act only during the treatment period. The mouse-skin assay has been used to establish a dose-response relationship for the induction of tumors by smoke particles.

Smoke particles have been fractionated into neutral, acidic, basic, and insoluble fractions. Fractionation has established that the tar subfractions contain the bulk of the polynuclear aromatic hydrocarbons (PAHs) and are the only substances that produce carcinomas. The subfractions of PAH contain neutral cocarcinogens that potentiate the carcinogenic PAHs. The weakly acidic fraction has been shown to contain promoters, as well as cocarcinogens; these include phenolic compounds and catechols.

Recent studies (reviewed in IARC, 1986) have demonstrated that tobacco smoke has transplacental carcinogenic effects in hamsters. Compounds that require metabolic activation to an active carcinogenic form—such as *N*-nitrosamine, benzo[a]pyrene, *o*-toluidine, ethylcarbamate, and vinyl chloride—act as transplacental carcinogens. Nitrosamine can also be formed in the fetus from unmetabolized nicotine.

More than 90% of the total weight of mainstream smoke is made up of vapor-phase components (i.e., material of which more than 50% passes through a Cambridge glass filter). Nonfiltered and conventional filtered cigarettes contain CO at 3–5 vol % per puff and 13–26 μg per cigarette. Cigar smoke contains CO at up to 11 vol %. Concentrations of many vapor-phase components in cigarette smoke vary directly with concentrations of tar and nicotine; that is not the case with nitrogen oxides (NO_x), most of which (95%) is nitric oxide. The NO_x content of an average American cigarette is 270–280 μg per cigarette.

Ciliotoxic agents inhibit lung clearance. Many of those agents are vapor-phase components of cigarette smoke; they include hydrocarbons (280–550 μg per cigarette), NO_2 (0–30 μg per cigarette), and formaldehyde (20–90 μg per cigarette). Hydrazines are effective carcinogens. Vinyl chloride is acted on by the cytochrome P-450 enzyme system to form a halogenated epoxide that can yield halogenated aldehydes or alcohols. The activated forms of vinyl chloride can react with adenosines or cytidines to form new rings that interfere with base pairings.

Cigarette-smokers have an increased risk of cancer in organs other than the lungs, such as the esophagus, the pancreas, and the urinary bladder. Cigarette smoke does not contact organs directly; therefore, mechanisms other than direct contact must be involved in carcinogenesis. The increased risk might be due to the formation of organ-specific carcinogens or to a shift in metabolism of common components toward localized formation of carcinogenic metabolites. Examples of organ-specific carcinogens are 2-naphthylamine, 4-aminobiphenyl, and benzidine. Volatile nitrosamines are present in tobacco and tobacco smoke; nitrosamines are also derived from residues of agricultural chemicals on tobacco.

The Ames bacterial assay has been used to identify and characterize mutagens in tobacco smoke. In that assay, mutagen content of smoke can be obtained from the equivalent of 0.01 cigarette, whereas only 0.0025 cigarette is needed for the sister-chromatid-exchange study. Substances with mutagenic potential are generally carcinogens as well. Activated carcinogens or mutagens form covalent adducts with cellular macromolecules. Those binding to DNA form bulky adducts that produce errors in DNA replication and promote mutation. The urine of cigarette-smokers contains 5–10 times the quantity of mutagens found in nonsmokers. Mutagen concentrations are highest in the evening and lowest in the morning. There appears to be no threshold in the dose-response relationship of smoking and lung-cancer incidence. That observation makes the incidental passive exposure of nonsmokers a potential mechanism of importance in the development of lung cancer in nonsmokers (IARC, 1986; NRC, 1986). Sidestream smoke contains larger quantities of *N*-nitrosamines and small particles. Cotinine in urine is a marker of cigarette-smoke exposure, and infants with care givers who smoke have been shown to excrete this material, which is evidence that the infants have been exposed to cigarette smoke.

The toxic components in cigarettes include simple diatomic gases that can have direct and long-term toxic effects, as well as compounds that are incomplete carcinogens and cocarcinogens and can produce their effects only after long periods of exposure (several decades). If we were faced with the question of finding *the* toxic component in a sample of cigarette smoke, without knowledge of how it had been generated or of the composition of the original material, a study of its "toxic component" might be extremely complex. Even after decades of serious investigation, we do not understand the role of tobacco-smoke components in producing chronic diseases, such as arteriosclerosis, emphysema, and malignant neoplasms. The task of identifying the toxic components is overwhelming and must be considered currently impossible. We can, however, identify groups of agents from a knowledge of their chemical similarity to agents generated in a standard control substance.

We can learn something about the potency of tobacco smoke by comparing the composition of tobaccos grown in different soils, after different periods of aging, after the use of various fertilizer additives, and after treatment with different methods, such as nicotine and tar extraction. The potency of each class of agents must be compared separately. The acute effects, chronic effects, and carcinogenic effects must be separately analyzed and compared with those of standard or control compounds in each group. The toxicologic evaluation of tobacco-smoke components might be both useful and confusing—useful as a means of understanding the effects of individual components, and confusing because the end product of analysis might not represent the active transient agent that causes an effect.

To be hazardous, cigarette smoke must be inhaled over long periods. However, acute effects can be produced in persons with hyperactive airways, such as persons with allergic asthma or bronchitis, or in persons with lung or heart diseases. Thus, the host conditions are important in determining risk. Recent work has demonstrated that not only the smoker, but also persons close by, can be affected by the smoke.

FIRE ATMOSPHERES

The thermal-degradation products that evolve when materials pyrolyze, smolder, or burn provide a prime example of complex mixtures known to incapacitate, injure, and cause death. The assessment of the toxicity of these complex mixtures has to begin with the determination of what is known about them. For example:

- All fire atmospheres are toxic.
- The main route of entry into the body is inhalation, although dermal absorption can play a role.
- Acute exposures (i.e., single exposures of less than 30 minutes) are the principal concerns (except for fire fighters, who receive chronic exposures).
- In most situations, the exposure concentrations are high; that is, one does not need to extrapolate results from high experimental doses to low doses that are considered to represent normal population exposures.

What is unknown stems from the fact that no two fires are alike. They differ in type and size of burning compartment, burning materials, and ventilation conditions. Each fire is a dynamic phenomenon that affects and is affected by the rapidly changing environmental conditions. Most fires involve multiple materials, each of which might produce hundreds of combustion products (Levin, 1986). The magnitude of the problem becomes apparent with the recognition that the toxicities of most of the products identified are unknown and that the number of toxic product combinations that could lead to possible interactions is immense.

Thus, posing the correct question to determine the best testing strategy is important. The question that initially concerned fire investigators was: Do any materials produce unusually toxic or extremely toxic thermal degradation products? ("Unusually toxic" implies that unexpected adverse effects are produced; "extremely toxic" implies that relatively small amounts of material produce measurable toxicity.) The testing strategy to answer these questions is obviously in the category of comparative toxic potency and is "effect-driven." The concern is not *why* a material produces an effect, but rather whether it produces the effect, how it ranks toxicologically in comparison with other materials that are suitable for the uses in question, and whether it produces an expected or unexpected effect.

In response to those questions, several test methods were designed (Alarie and Anderson, 1979; Crane et al., 1977; Hilado, 1976; Klimisch et al., 1980; Levin et al., 1982; Yusa, 1985). Although different, these methods have some similarities. For example, all include the testing of complete samples and measurement of the acute inhalation toxicity in experimental animals. Complete chemical analysis of combustion products was not considered the proper approach, because toxic components might be missed, either through lack of detection or through lack of knowledge of their toxicity. Animals, however, would breathe a combination of the gases and respond to its toxicity, regardless of whether it was due to one component or the interaction of several components. To avoid false-negative results, the tests must be designed to simulate real fire conditions or worst-case conditions. That approach provides a measure of the relative intrinsic toxicity of the atmospheres produced by materials that are thermally decomposed under the specified conditions of each test method. It can be used by material manufacturers or product developers who need to choose materials to market.

The comparative toxic-potency approach, however, is scientifically unsatisfactory, because it fails to provide any clues as to which components are responsible for the toxicity and makes it nearly impossible to predict toxicity of future samples. The testing strategy to determine the causative agents in such mixtures therefore involves extensive chemical analysis. The analysis must be preceded by toxicity testing of the complete mixture to indicate whether there is a cause for concern. The effect-driven chemical analysis was used to determine that a fire-retarded laboratory formulation of rigid polyurethane foam produced a highly toxic bicyclic phosphate ester when thermally decomposed (Petajan et al., 1975).

Thus, the testing strategy used to determine the toxicity of the thermal-degradation products of materials has historically consisted of tests of comparative toxic potency combined with effect-driven chemical analysis. That approach, however, has raised even more questions. For example, do materials with minor differences between formulations or dye lots, or with minor changes in additives or fire retardants, and so on, all need to be tested? If the components of materials have been tested, do composites made of the components need to be tested? What additional hazard is generated when a new material is added to a room full of furniture? Toxicity is only one of many kinds of information needed for a hazard assessment, and it is the hazard that must be considered when one is making decisions on the suitability of material for a particular use. Other information needed for a hazard assessment includes the quantity of material present, its configuration, the proximity of other combustible materials, the volume of the compartments to which combustion products can spread, ventilation conditions, ignition and combustion properties of materials present, the presence of ignition sources, the presence of fire protection systems, and the building occupancy.

Determination of toxicity for a hazard assessment requires a testing strategy different from that needed for the relative ranking of materials. One needs to develop a model to predict toxicity while avoiding expensive and time-consuming multiple tests. An approach that examines the toxicity of both the primary component (individually and combined) and the total mixture has been used (Levin et al., in press). To use that single- and mixed-component approach, toxicity data on the primary toxic gases present in most fires were needed. Research began on a few of these gases (CO, CO_2, and HCN) and on the effect of oxygen concentration to determine their individual toxicities and the occurrence of interactions. It was demonstrated in rats exposed to CO and HCN for 30 minutes at various concentrations that these two gases act in an additive fashion, such that some animals will die if the sum of [CO]/LC_{50}CO and [HCN]/LC_{50}HCN is approximately equal to 1 ± 15%. If the sum is greater or less than that value, all the animals should die or live, respectively. It has also been shown in 30-minute exposures that CO_2 acts synergistically with CO; in the presence of approximately 5% CO_2, the toxicity of CO (as indicated by lethality) doubles (Figure C-1) (Levin et al., 1987). Relationships among CO, CO_2, and HCN were also determined, as was the effect of O_2 deprivation. (O_2 deprivation is probably more of a problem in the room of fire origin than at any distance away from the fire.) The experimental design was based on a matrix that indicated what concentrations and combinations of gases were lethal (Figure C-2). With that matrix, a testing strategy that uses single and combined primary gases can be applied to predict toxicity as follows. First, a material is

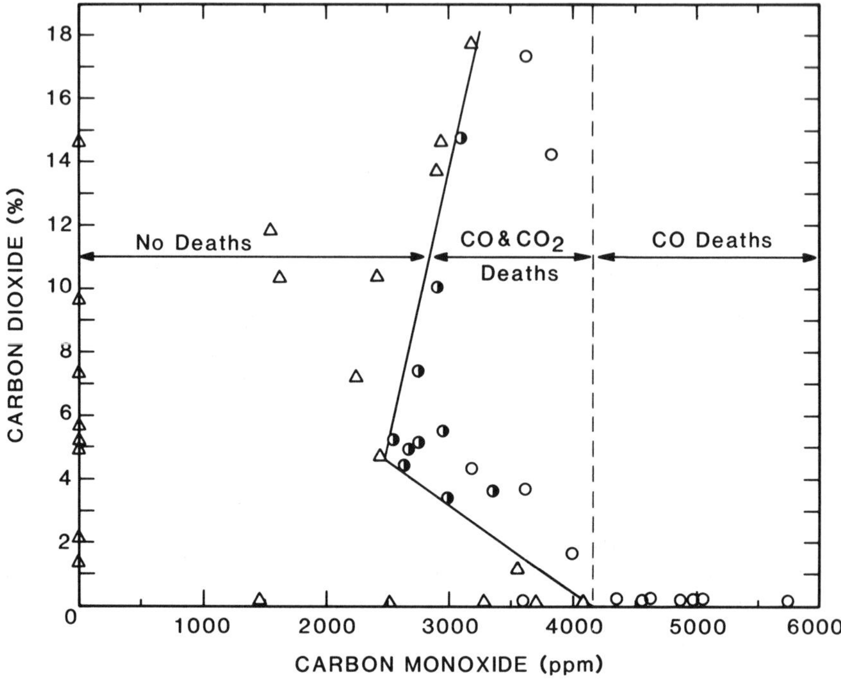

FIGURE C-1 Deaths resulting from exposure to CO or CO_2 alone and from CO plus CO_2. No deaths (△); deaths during exposure (○); deaths during and after exposure (◑). Solid line separates experiments in which no deaths occurred from those in which one or more animals died.

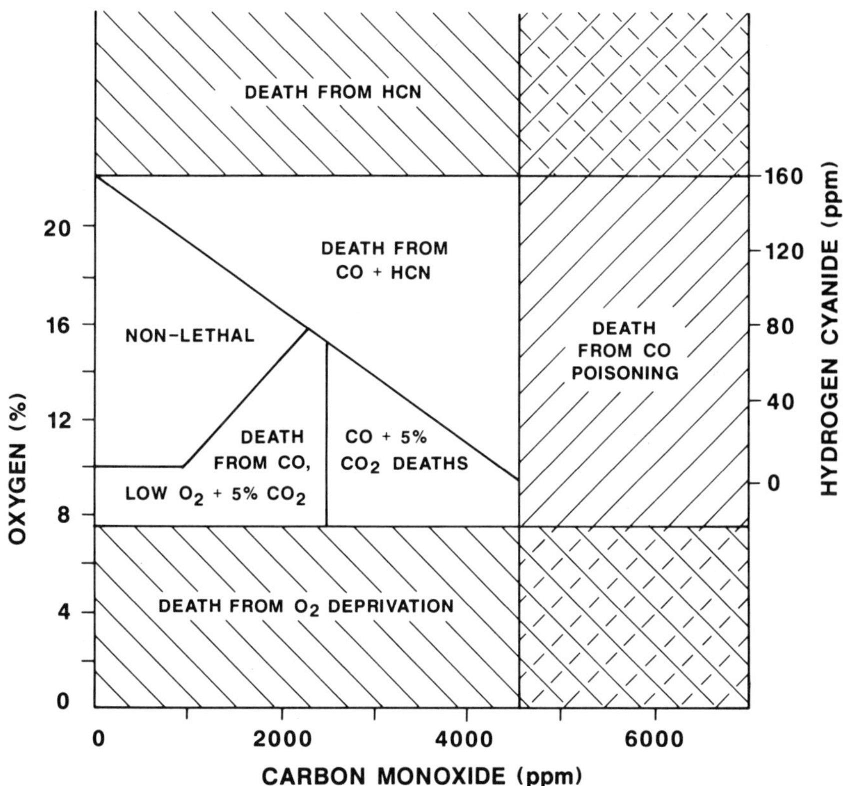

FIGURE C-2 Lethal concentrations of CO, O_2, and HCN individually and in combination with 5% CO_2.

thermally decomposed, and the concentrations of the primary gases are measured. An LC_{50} of each is then predicted on the basis of the previously determined relationships between the gases. To test the validity of the prediction, animals are exposed at the predicted LC_{50}. If some fraction of the exposed animals dies, it can be assumed that the LC_{50} has been estimated fairly closely. (That is possible in combustion toxicology, because concentration-response curves are usually very steep.) If none of or all the animals die, more experiments are conducted to determine the LC_{50} more precisely. If the material produces combustion products that are significantly more toxic than was predicted, according to the pure-gas relationships, further studies are necessary to determine what other toxic products or interactions have occurred. As more data are collected on materials, more gases will be added to the model, and the predictions will become more accurate.

HEXACARBON NEUROPATHY

Hexacarbon neuropathy is a relatively newly discovered clinical disorder. The traditional belief was that aliphatic hydrocarbons have no organ-specific activity and are inert metabolically. The emergence of hexacarbon neuropathy established that the belief had no foundation. The investigation of the disease and the unraveling of the causative agent(s) provide an illustration of strategies for testing the toxicity of complex mixtures. The following description is based on three published reviews of the topic (Allen, 1980; Couri and Milks, 1985; Scala, 1976).

The initial reports of hexacarbon neuropathy were associated with exposure to "*n*-hexane" (composition unknown). They arose in Japan in 1964, and the first American cases were reported in 1971. A large outbreak of peripheral neuropathy took place in 1973 in an Ohio coating plant. By 1976, the American episodes had been linked to a possible common neurotoxic metabolite formed from the suspect agents. In each case, despite some initial "identification" of an active agent, there was considerable uncertainty. There was, for example, an unwillingness to believe that otherwise inert agents or small changes in formulation could produce the reported effects in humans. The index cases of hexane neuropathy in the United States were ascribed to *n*-hexane, but the disbelief of some investigators led to chemical analyses. A sample of solvent from the workplace was found to contain 68% acetone, 15% isomeric C_6 and C_7 hydrocarbons, and only 17% *n*-hexane. Other case reports from outside the United States used solvents that contained a variety of paraffinic, cycloparaffinic, and aromatic hydrocarbons and occasionally some esters. The causative agent was determined through a form of bioassay-directed fractionation. Various hydrocarbon solvents were tested in animals to determine whether *n*-hexane was the active agent, whether any other hexane or heptane isomers were active, and whether some oxygenates could increase the effect. The bioassay system was validated with respect to producing clinical or histologic evidence of neuropathy. The potential of *n*-hexane alone to produce neuropathy was well established. The final link was to create a synthetic blend that simulated commercial hexanes, but contained no *n*-hexane. The final experiment (Egan et al., 1980), not cited in the major reviews referred to earlier, showed that both "*n*-hexane-free hexane mixture" and the possibly interacting agent methyl ethyl ketone fail to produce the characteristic neuropathy.

The second mystery involving hexacarbons yielded in part to a form of screening study. The inquiry was guided to some degree by the circumstances surrounding uncovery of the disease. The peripheral-neuropathy "outbreak" in Ohio in 1973 arose from an index case in a person who complained of progressive weakness. That case generated the names of five co-workers with similar complaints who had been seen individually elsewhere. Before the investigation was completed, over 1,000 workers had been studied; 86 had some manifestation of a peripheral neuropathy ultimately related to workplace exposures. Their work involved the production of vinyl-coated fabrics and vinyl sheets for coverings. Work practices in the plant involved extensive dermal and inhalation exposure to many chemicals. The investigative team narrowed the 275 chemicals received by the plant to the following suspected neurotoxins: methyl ethyl ketone, methyl *n*-butyl ketone, *n*-hexane, toluene, acrylamide, phthalate esters, organophosphates, furans, heavy metals, carbon disulfide, methanol, and trichloroethylene.

The clinical pattern of the disease (mixed sensory and motor effects with symmetrical distribution in distal portions of the limbs, without abnormalities elsewhere in the nervous system or the skin) tended to rule out heavy metals, methanol, and trichloroethylene quickly. The findings were similar to those reported for *n*-hexane and one case of neuropathy reported for methyl *n*-butyl ketone. Clinical chemistry studies ruled out organophosphates. Epidemiologic considerations had focused attention on the print department. Workplace measurements tended to rule out methyl isobutyl ketone, *n*-hexane, toluene, acetone, and some other materials, but indicated substantial exposures to methyl *n*-butyl ketone and methyl ethyl ketone. The time course of the disease coincided with the substitution of methyl *n*-butyl ketone for methyl isobutyl ketone in an ink thinner and solvent system. In a fortunate coincidence, it was found that another plant that used the same processes but not methyl *n*-butyl ketone had no cases of neuropathy. Animal studies showed that methyl ethyl ketone and methyl isobutyl ketone did not produce neuropathy, but that methyl *n*-butyl ketone did. It is now generally held that methyl *n*-butyl ketone was the causative agent in this outbreak.

By 1976, metabolic studies were far enough advanced at least to suggest that the 1971 *n*-hexane outbreak and the 1973 methyl *n*-butyl ketone cases were linked through a common metabolite, 2,5-hexanedione. It has since been shown that omega-1 oxidation of C_6 hydrocarbons or oxygenates to a gamma diketone yields active neurotoxicants.

The hexacarbon-neuropathy case studies reveal that an organized, scientific approach to problem-solving can succeed, even with complex mixtures, where there is a combination of focused questions, assay methods, and good luck. The focused questions resulted in the isolation of the problem to specific work areas that used materials (mixtures) of known or knowable composition. The same focused questions went beyond a general description of the disease to its specific characteristics and permitted the establishment of links to other case reports. The availability of suitable assay systems allowed, in the 1973 episode, the screening of hundreds of workers for neurologic deficits and permitted the concerns in the workplace to be concentrated in the print department. The assays also allowed some agents to be eliminated as causes, because the neurologic findings were not of the proper sort. An animal assay allowed mixtures or individual compounds to be screened for the neurologic disorder in question and structure-activity studies to be conducted.

With respect to the interests of this committee, the testing strategies used in evaluating the complex mixtures were largely effect-driven. Agents and workers were screened for a definite effect. The success of the screening was evident, but was only a portion of the larger success in solving two related industrial-health problems.

REFERENCES

Alarie, Y. C., and R. C. Anderson. 1979. Toxicologic and acute lethal hazard evaluation of thermal decomposition products of synthetic and natural polymers. Toxicol. Appl. Pharmacol. 51:341–362.

Allen, N. 1980. Identification of methyl *n*-butyl ketone as the causative agent, pp. 834–845. In P. S. Spencer and H. H. Schaumburg (eds.). Experimental and Clinical Neurotoxicology. Williams & Wilkins, Baltimore.

Couri, D., and M. M. Milks. 1985. Hexacarbon neuropathy: Tracking a toxin. Neurotoxicology 6(4):65–71.

Crane, C. R., D. C. Sanders, B. R. Endecott, J. K. Abbott, and P. W. Smith. 1977. Inhalation Toxicology. I. Design of a Small Animal Test System. II. Determination of the Relative Toxic Hazards of 75 Aircraft Cabin Materials. FAA-AM-77-9. U.S. Federal Aviation Administration, Office of Aviation Medicine, Washington, D.C. (49 pp.)

Egan, G., P. Spencer, H. Schaumburg, K. J. Murray, M. Bischoff, and R. Scala. 1980. *n*-Hexane-"free" hexane mixture fails to produce nervous system damage. Neurotoxicology 1:515–524.

Hilado, C. J. 1976. Relative toxicity of pyrolysis products of some foams and fabrics. J. Combust. Toxicol. 3:32–60.

IARC (International Agency for Research on Cancer). 1986. IARC Monographs on the Evaluation of the Carcinogenic Risk of Chemicals to Humans. Vol. 38. Tobacco Smoking. International Agency for Research on Cancer, Lyon, France. (421 pp.)

Klimisch, H.-J., H. W. M. Hollander, and J. Thyssen. 1980. Comparative measurements of the toxicity to laboratory animals of products of thermal decomposition generated by the method of DIN 53 436. J. Combust. Toxicol. 7:209–230.

Levin, B. C. 1986. A Summary of the NBS Literature Reviews on the Chemical Nature and Toxicity of the Pyrolysis and Combustion Products from Seven Plastics: Acrylonitrile-Butadiene-Styrenes (ABS), Nylons, Polyesters, Polyethylenes, Polystyrenes, Poly(Vinyl Chlorides) and Rigid Polyurethane Foams. NBSIR 85-3267. National Bureau of Standards, Gaithersburg, Md.

Levin, B. C., A. J. Fowell, M. M. Birky, M. Paabo, A. Stolte, and D. Malek. 1982. Further Development of a Test Method for the Assessment of the Acute Inhalation Toxicity of Combustion Products. NBSIR 82-2532. National Bureau of Standards, Gaithersburg, Md.

Levin, B., M. Paabo, J. L. Gurman, and S. E. Harris. In press. Effects of exposure to single or multiple combinations of the predominant toxic gases and low oxygen atmospheres produced in fires. Fundam. Appl. Toxicol.

NRC (National Research Council). 1986. Environmental Tobacco Smoke: Measuring Exposures and Assessing Health Effects. National Academy Press, Washington, D.C. (337 pp.)

Petajan, J. H., K. J. Voorhees, S. C. Packham, R. G. Baldwin, I. N. Einhorn, M. L. Grunnet, B. G. Dinger, and M. M. Birky. 1975. Extreme toxicity from combustion products of a fire-retarded polyurethane foam. Science 187:742–744.

Scala, R. A. 1976. Hydrocarbon neuropathy. Ann. Occup. Hyg. 19:293–299.

Yusa, S. 1985. Development of laboratory test apparatus for evaluation of toxicity of combustion products of materials in fire, pp. 471–487. In 7th Joint Meeting of the U.S.-Japan Panel on Fire Research and Safety, Washington, D.C., October 24–28, 1983. Proceedings. NBSIR-85/3118. National Bureau of Standards, Gaithersburg, Md. (Available from NTIS as PB85-199545/XAB.)

D

Predicting the Joint Risk of a Mixture in Terms of the Component Risks

One type of complex-mixture problem is illustrated by the following paradigm. One has a complex mixture of known chemical composition, the toxicities of whose individual components can be estimated, and the overall toxicity of the mixture needs to be estimated. At times, information on the toxicity of the mixture has been obtained from a rodent experiment in which the mixture has been tested at high dosages, or epidemiologic data have been gathered on the mixture or on similar mixtures. However, usually the only available toxicity information on a mixture is related to its individual components, and one needs to estimate the toxicity of the mixture.

This appendix considers the problem of estimating the joint risk of a mixture in terms of the component risks—that is, in terms of the risks of the individual chemicals in the mixture. It begins with a class of models in which the joint risk can be expressed in terms of the component risks. Thus, if one had reason to believe in the validity of a particular model within this class and had information on the component toxicities, the joint risk could be estimated. It then discusses an approach when there is no preferred model, but there is some confidence that the true model may be within the class of models being considered. The approach is analogous, in some respects, to one sometimes used to assess the carcinogenic risk associated with a chemical on the basis of laboratory data—that is, the use of several models (e.g., one-hit, multistage, multihit, Weibull, and probit models) to predict the risks associated with small exposures. The resulting range of risks is presented as a "plausible" range. In the present situation, we adapt that idea to estimate the risk associated with a mixture. Specifically, we examine a variety of generalized additive models for the joint risk associated with the mixture and express the risk for each model in terms of the risks associated with the individual components.

Of course, one usually cannot discriminate between various models for complex mixtures at low doses, because adequate response data are generally not available. However, we hope that the range of predicted risks generated from several of the commonly used models will provide a plausible range of risks. The plausible range might be used in several ways, including application in a sensitivity analysis of estimates based on a specific model.

Before the idea is described in greater detail, several cautions are in order. First, the results given below are preliminary and are intended mainly to illustrate the idea; full development and evaluation of the approach would require additional research, which, although perhaps not extensive, is nevertheless lacking. Second, the approach is not intended to be able to identify the true risk associated with a mixture at low doses—that is unverifiable. Rather, we want to produce an estimate of the low-dose risk that is consistent with a class of models of the joint effects of the mixture.

EXPRESSING JOINT RISKS IN TERMS OF COMPONENT RISKS

Consider the multistage model for two compounds that is developed in Appendix E, in which it is assumed that no two components affect the same transition. Let $P(x_1, x_2)$ denote the risk (i.e., probability of developing a tumor) associated with exposure to doses x_1 and x_2 of the two compounds. Then, for this form of multistage model,

$$P(x_1, x_2) = 1 - \exp[-A(1 + B_1x_1)(1 + B_2x_2)]. \qquad \text{(D-1)}$$

The background risk is $P(0,0) = 1 - \exp[-A]$, and the risks associated with the individual components are $P(x_1,0) = 1 - \exp[-A(1 + B_1x_1)]$ and $P(0,x_2) = 1 - \exp[-A(1 + B_2x_2)]$. We will demonstrate that the combined risk $P(x_1, x_2)$ can be determined by knowing $P(0,0)$, $P(x_1,0)$, and $P(0,x_2)$. Consider the function

$$g(p) = \ln[-\ln(1 - p)].$$

By algebraic manipulation, it can be verified that Equation D-1 can be equivalently expressed as:

$$P(x_1, x_2) = 1 - \exp\{-\exp[g(P[x_1,0]) + g[P(0,x_2)] - g[P(0,0)]\}.$$

With Equation D-1, the joint risk, $P(x_1, x_2)$, can be completely specified from the component and background risks.

The above result can be extended to any class of *generalized additive models* that have the form

$$F[P(x_1, x_2)] = B_0 + B_1z_1(x_1) + B_2z_2(x_2), \qquad \text{(D-2)}$$

where z_1 and z_2 are any transformations of x_1 and x_2 and F is any invertible function. Without loss of generality, we can assume that $z_1(0) = z_2(0) = 0$. The background and component risks must then satisfy

$$F[P(0,0)] = B_0,$$
$$F[P(x_1,0)] = B_0 + B_1 z_1 (x_1),$$

and

$$F[P(0,x_2)] = B_0 + B_2 z_2(x_2).$$

Thus, we can express Equation D-2 as

$$\begin{aligned} F[P(x_1, x_2)] &= F[P(0,0)] + \{F[P(x_1,0)] - F[P(0,0)]\} \\ &\quad + \{F[P(0,x_2)] - F[P(0,0)]\} \\ &= F[P(x_1,0)] + F[P(0,x_2)] - F[P(0,0)], \end{aligned}$$

and can write

$$P(x_1, x_2) = F - \{F[P(x_1, 0)] + F[P(0,x_2)] - F[P(0,0)]\}.$$

Hence, $P(x_1, x_2)$ is determinable from the background and component risks.

The class of models that satisfy Equation D-2 is broad and includes the linear, multiplicative, logistic, and multistage models as subclasses. We illustrate those in the following table, where we use x_i^* to denote $z_i(x_i)$.

We also note that, for a multiplicative model, the relative risk of the mixture—say, $RR(x_1, x_2)$—can be expressed as the product of the component relative risks; that is,

$$RR(x_1, x_2) = [RR(x_1,0)]\,[RR(0,x_2)]. \qquad \text{(D-3)}$$

That result is useful for situations where only the relative risks of the component chemicals are available (and not their actual risks). Moreover, for sufficiently small x_1 and x_2, the above result also applies to the multistage model given above and the logistic model.

ILLUSTRATIVE EXAMPLES

Reif (1984) gave several examples of the collection of epidemiologic data on both the individual and the joint effects of two "substances," one of which is tobacco-smoking. We use the data to illustrate the multiplicative and logistic models. For example, Reif examined the risk of lung cancer as a function of smoking status and uranium exposure (see Table D-2). Letting A = smoking and B = uranium, he got:

$$p^* = 1.5/26{,}392 = 0.57 \times 10^{-4},$$
$$P(x_1, 0) = 5.87 \times 10^{-4},$$
$$P(0, x_2) = 2.27 \times 10^{-4}, \text{ and}$$
$$P(x_1, x_2) = 22.7 \times 10^{-4}.$$

TABLE D-1 Alternate Response Models

Model	Formula for $P(x_1,x_2)$	Form of $F(p)$
Linear	$B_0 + B_1x_1^* + B_2x_2^*$	p
Multiplicative	$\exp[B_0 + B_1x_1^* + B_2x_2^*]$	$\log p$
Logistic	$\dfrac{\exp[B_0 + B_1x_1^* + B_2x_2^*]}{1 + \exp[B_0 + B_1x_1^* + B_2x_2^*]}$	$\log[p/1 - P)]$
Multistage	$1 - \exp[-B_0(1 + B_1x_1^*)(1 + B_2x_2^*)]$	$\log[-\log(1 - p)]$

The estimate of $P(x_1, x_2)$ based on the multiplicative model (see Table D-1) is 23.4×10^{-4}. The logistic and the multistage models give very similar estimates. Thus, in this example, all three models give very similar estimates of $P(x_1, x_2)$, and ones that closely agree with the value $P(x_1, x_2)$ estimated by mixture data. With each model, we can predict $P(x_1, x_2)$ accurately with only the information on individual cancer risks.

Reif gave five other examples (his Tables 2-6). We summarize the above results and those for the other examples in Table E-2.

The model-based estimates of $P(x_1, x_2)$ for the first three examples, in which there are data on $P(0,0)$, $P(x_1,0)$, and $P(0,x_2)$, are quite good. The three models give similar predictions, all closely approximating the $P(x_1,x_2)$ estimated directly. The last three examples are from case-control studies in which only relative risks are available (and not estimates of $P(0,0)$, $P(x_1,0)$, and $P(0, x_2)$. Thus, we can predict only the relative risk of the mixture on the basis of the multiplicative model. This estimate closely approximates $RR(x_1,x_2)$ for the fourth example, but underestimates it for the fifth and sixth examples. The results of the multistage model might seem to be inconsistent with those described in Appendix E, where it is argued that the excess risk associated with a mixture (at low exposures) can often be accurately approximated as the sum of the individual excess risks. However, on the basis of Reif's data (which reflect relatively high exposures), the excess risks associated with mixtures are consistently higher than the sums of the individual excess risks. Moreover, evaluation of the multistage model for the data produces a joint risk estimate that exceeds the sum of the individual estimates. Thus, whereas the multistage model gives good predictions of the joint risk in the examples, the exposures were not so low that it approximates a linear model. These results verify the critical dependence of the argument in Appendix E on magnitude of exposure. The risk estimates derived from the various epidemiologic studies are also subject to considerable uncertainty.

MIXTURES OF MORE THAN TWO CHEMICALS

The results extend in a straightforward way to the situation of mixtures of K compounds. The analogue of Equation D-2 is

TABLE D-2 Illustrative Examples Based on Epidemiology Data

Table in Reif (1984)	End Point	A	B	p^*	$P(x_1,0)^a$	$P(0,x_2)^a$	True $P(x_1,x_2)$	Estimates of $P(x_1,x_2)$ Based On: Multiplicative Model	Logistic Model	Multistage Model
1	Lung cancer	Smoking	Uranium mining	0.57×10^{-4}	5.87×10^{-4}	2.27×10^{-4}	22.7×10^{-4}	23.4×10^{-4}	23.4×10^{-4}	23.2×10^{-4}
2	Abnormal sputum	Smoking	Uranium mining	0.04	0.11	0.08	0.22	0.22	0.21	0.21
3	Lung cancer	Smoking	Asbestos workers	1.13×10^{-7}	12.3×10^{-7}	5.84×10^{-7}	60.2×10^{-7}	63.6×10^{-7}	63.6×10^{-7}	63.9×10^{-7}
4[b]	Lung cancer	Smoking	Asbestos miners	1.0	12.0	1.6	19.0	19.2	*c*	*c*
5[b]	Oral cancer	Smoking	Alcohol use	1.0	1.53	1.23	5.71	1.9	*c*	*c*
6[b]	Renal cancer	Smoking	Cadmium	1.0	1.0	0.8	4.4	0.8	*c*	*c*

[a] These probabilities are best understood as defined in a relative sense; if converted to lifetime risk for exposed population, they reflect high-dose phenomena.
[b] Data based on relative risks. Entries for $P(x_1,0)$, $P(0,x_2)$, and $P(x_1,x_2)$ denote relative risks. Risk data, RR_{AB}, are approximated by Equation D-3.
[c] These estimates cannot be computed only from information on the *relative* risks of components.

$$F[P(x_1, x_2, \ldots, x_k)] = B_0 + B_1 z_1(x_1) + \ldots + B_k z_k(x_k). \quad \text{(D-4)}$$

For example, the logistic model (see Table D-1) generalizes to:

$$P(x_1, x_2, \ldots, x_k) = P_1 P_2 \ldots P_k/[p^{*(k-1)}]$$

where $P_j = P(0, 0, \ldots, x_j, 0, \ldots, 0)$ for $j = 1$ to k—the effect with exposure to only substance j (at dose x_j).

ADDITIONAL RESEARCH

For a proper assessment of the potential of this approach, several problems need to be investigated. First, a variety of models should be considered, and the formula for $P(x_1,x_2)$ for each (in terms of the individual effects) should be derived for each model. Then, when information on individual and joint effects is available, all the estimates should be computed with the models. Given the results for several data sets, some models might be deemed more realistic, because they yield results that are consistent with what the real risk is thought to be.

The range of plausible values of $P(x_1,x_2)$ can also be applied to data sets in which there is no direct information on $P(x_1,x_2)$, as an aid in designing (or deciding to undertake) an extensive experiment. For example, if the plausible range is within achievable or acceptable limits, it might be decided that little would be gained from carrying out an extensive experiment.

SUMMARY

Generalized additive models are discussed more specifically in the context of a situation in which risk estimates associated with individual components of a mixture are available. Here, one goal is to be able to express the risk associated with the mixture in terms of the component risks, so that, for a given model, the former can be estimated in terms of the latter. Another use of the result is to predict a plausible range of risks associated with the mixture.

REFERENCE

Reif, A. E. 1984. Synergism in carcinogenesis. J. Natl. Cancer Inst. 73:25–39.

E

Cancer Models

One ultimate purpose of laboratory investigations of potential carcinogens is to predict the human health effects of mixtures to which the population is exposed environmentally. Mathematical models have been applied throughout the history of toxicity testing to describe dose-response relationships that might allow extrapolating from the high doses used in the laboratory to the low doses generally encountered in the environment. Understanding of the biologic mechanisms of carcinogenesis has advanced, and newer models describe dose-response relationships in a manner that reflects newer knowledge of the biologic bases of disease.

When the carcinogenic effects of a mixture, instead of single chemicals, are being considered, the concern is usually over whether interactions might occur between components of the mixture and yield toxicity greater than what would be expected, given knowledge of the toxicity of the individual materials. Models of interaction observed at the high doses used in laboratory bioassays of carcinogenicity might predict additivity at lower, environmental doses. The purpose of this appendix is to detail and discuss that phenomenon. When mixture components are tested for carcinogenicity in a laboratory bioassay, inclusion of high-dose multiple-exposure groups might yield little useful information, if human environmental exposure to the mixture occurs only at low doses. In such a case, the results of this appendix suggest that cancer predicted from bioassays of the individual mixture components can yield a good estimate of the cancer risk associated with human exposure to the mixture. The basis for and qualification of that conclusion are elaborated in the next section.

IMPLICATIONS OF THE MULTISTAGE MODEL

One approach to estimating the cancer risk associated with exposure to multiple carcinogenic agents at low doses is to assume that the excess risks are additive, that is, that the excess risk of the mixture is equal to the sum of the excess risks of the components.

The multistage model, originally developed by Berenblum-Shubik (1949) and Armitage and Doll (1954), provides a basic theoretical framework that can be extended to account for exposure to multiple agents. The following section describes how the multistage model can be generalized to take into account exposure to multiple agents. The generalized model is used to analyze the relationship between the extent of interaction observed in a bioassay and the departure from additivity at environmental exposures. On the basis of this examination, general principles concerning the most efficient approaches for estimating the joint effects of multiple agents at environmental doses are discussed.

Derivation of the Multistage Model that Accounts for Exposure to Multiple Carcinogenic Agents

The multistage model assumes that a cell becomes a cancer cell by progressing through an ordered sequence of stages. The number of stages is assumed to be some small number k, and the cell is neoplastic once it completes the kth stage.

Following the approach taken by Whittemore and Keller (1978), we define

$P_i(t)$ = probability that a cell is in the ith stage at time t, the age of the organism, and

$\lambda_i(t)$ = transition rate from i to the ($i^{\text{th}} + 1$) stage at time t.

The product of these two terms, $[\lambda_i(t)][P_i(t)]$, is the instantaneous probability that a cell leaves the ith stage for the ($i^{\text{th}} + 1$) stage at time t. The rate of transition out of the ith stage at time t is defined as $dP_i(t)/dt$. This instantaneous rate of transition is equal to the instantaneous probability of transformation of a cell from the ($i^{\text{th}} - 1$) to the ith stage minus the instantaneous probability of a transformation from the ith to the ($i^{\text{th}} + 1$) stage. This set of conditions defines the set of simple differential equations:

$$\begin{aligned} dP_0(t)/dt &= [-\lambda_0(t)][P_0(t)] & P_0(0) &= 1 \\ dP_i(t)/dt &= [-\lambda_i(t)][P_i(t)] + [\lambda_{i-1}(t)][P_{i-1}(t)] & P_i(0) &= 0, i = 1, \ldots, k-1, \\ dP_k(t)/dt &= [\lambda_{k-1}(t)][P_{k-1}(t)] & P_k(0) &= 0, \end{aligned} \qquad \text{(E-1)}$$

which can be be solved to obtain $P_i(t)$.

To account for exposure to m carcinogenic agents, we assume the transition rate of a cell from the ith to the ($i + 1$) stage is

$$\lambda_i(t) = \alpha_i + \sum_{j=1}^{m} [\beta_{ij} x_j(t)]. \quad \text{(E-2)}$$

where $x_j(t)$ = exposure to j^{th} agent at time t,
α_i = background transition rate for i^{th} stage, and
β_{ij} = unit exposure transition rate for j^{th} agent on i^{th} stage.

Equation E-2 assumes that each of the molecules of the different agents in the mixture acts independently of the others with regard to the probability of causing a cell transformation.

Suppose we also assume the following:

- The time required for a cell in its kth or final, malignant transformed state to grow into a death-causing tumor is approximately constant and equal to w.
- The probability that a given individual cell causes a tumor is very small.
- An organ contains N cells of a specified type, each capable of being the origin of the tumor causing death.
- N is very large.
- Each of the cells acts independently with regard to becoming transformed and ultimately leading to a tumor.

Then the age-specific death rate due to a specified tumor type in a particular organ can be expressed, to a close approximation, as

$$h(t) = N \frac{dP_k(t - w)}{dt} = N[\lambda_{k-1}(t - w)][P_{k-1}(t - w)], \quad \text{(E-3)}$$

which is the instantaneous conditional probability that deaths from tumor will occur at time t given that the animal is alive just before time t. The unconditional probability of death due to that tumor by age t in the absence of competing risks is

$$P(t) = 1 - \exp\left\{- \int_0^t h(v)dv\right\}$$

$$= 1 - \exp\left\{- \int_w^t N[dP_k(v - w)/dv]dv\right\}. \quad \text{(E-4)}$$

We assume that exposures are constant and continuous over a lifetime, as is often the case in chronic carcinogenesis experiments. That means that $x_j(t) = x_j, j = 1, 2, \ldots, m$ and that the transition rates are constants. In this restricted case, Equation E-2 has the simplified form,

$$\lambda_i(t) = \lambda_i = [\alpha_i + \sum_{j=1}^{m} \beta_{ij} x_j]. \quad \text{(E-5)}$$

Note that this assumes that the exposure level x_j at time t determines $\lambda(t)$, as opposed to some other value as a part of the exposure history, such as cumulative exposure. Substituting this definition for the ith-stage transition rate in Equation E-1 and integrating both sides of the equation yields the iterative solution,

$$P_1(t) = \int_0^t \lambda_0 \, dv = \lambda_0 t$$

$$P_2(t) = \int_0^t \lambda_1 \lambda_0 \, v dv = \lambda_1 \lambda_0 \frac{t^2}{2!}$$

$$\vdots \qquad \text{(E-6)}$$

$$P_k(t) = \int_0^t \lambda_{k-1} \lambda_{k-2} \ldots \lambda_0 \frac{v^{k-1}}{(k-1)!} \, dv = \prod_{j=0}^{k-1} \lambda_j \frac{t^k}{k!}.$$

With these results, the age-specific death rate defined in Equation E-3 and the probability of death due to that tumor defined in Equation E-4 are expressible as

$$h(t) = N \frac{t^{k-1}}{(k-1)!} \prod_{j=0}^{k-1} \lambda_j \text{ and} \qquad \text{(E-7)}$$

$$P(t) = 1 - \exp\left\{-N \frac{t^k}{k!} \prod_{j=0}^{k-1} \lambda_j\right\}, \qquad \text{(E-8)}$$

respectively, where we have assumed $w = 0$ (for simplicity).

The next section looks at the simplest model of this form that shows a synergistic effect between carcinogens. A hypothetical biologic example is then described in which the parameters of the model are given a specific numerical form, so that we can investigate the effect of the synergism observed in a carcinogenesis bioassay on environmental-exposure cancer-risk estimates.

ILLUSTRATIVE MULTISTAGE MODEL THAT RESULTS IN SYNERGISTIC EFFECT

A simple multistage model that results in an effect greater than additive arises from the assumption that each of two carcinogenic agents affects the transition rates of different single stages in the multistage process. In the case of two agents ($m = 2$), where the first agent affects the ith stage and the second agent the jth stage and no other transition rates are affected, it follows from Equation E-5 that

$$\begin{aligned} \lambda_s &= \alpha_s && s \neq i, j \\ \lambda_i &= \alpha_i + \beta_{i1} x_1, \\ \lambda_j &= \alpha_j + \beta_{j2} x_2. \end{aligned} \tag{E-9}$$

Substituting these values into Equation E-8 yields the expression

$$P(t) = 1 - \exp\left\{-Nt^k \left[\prod_{s=0}^{k-1} \alpha_s\right](\alpha_i + \beta_{i1}x_1)(\alpha_j + \beta_{j2}x_2)/\alpha_i\alpha_j k!\right\}, \tag{E-10}$$

which may be written in the more concise form

$$P(t) = 1 - \exp\{-A(1 + B_1x_1)(1 + B_2x_2)\}, \tag{E-11}$$

where

$$A = Nt^k \prod_{s=0}^{k-1} \alpha_s/k!, \text{ and}$$

$B_1 = \beta_{i1}/\alpha_i$ and $B_2 = \beta_{j2}/\alpha_j$ are the relative transition rates for the first and second agents, respectively.

If the risk of death from nontumor causes is negligible to time t, Equation E-11 represents the probability of a tumor death by time t. Using this model, the next section describes the dose dependence of the synergistic effect.

Estimation of Largest Synergistic Effect Detectable in 2 × 2 Balanced-Design Experiment

The experimental design most often used for estimating the joint effects of multiple agents is the balanced 2 × 2 form consisting of four experimental groups: control, a single exposure group for each of two agents, and a joint exposure group at the same doses of the single agents. With that type of design, it is shown how large an interaction effect can be measured in a practical bioassay under the assumption that the dose-response relationship of two agents can be represented by the specific form of the generalized multistage model depicted in Equation E-11. That degree of interaction will represent a practical upper bound on the interaction of two agents at experimental doses, because greater interaction cannot be measured within the constraints of the chosen experimental design. On the basis of this degree of interaction, we assess the rate of error that is introduced by assuming that the joint risk is additive at environmental exposures. The approach establishes an error rate for additivity that is as great as can be directly measured from a 2 × 2 bioassay design defined by the example under the assumption that the multistage model in its generalized form is a valid representation of the joint carcinogenic effect of two agents. In the example that follows, it is assumed that there are 50 animals per exposure group.

To obtain an extreme interaction effect, we impose several additional conditions on our hypothetical bioassay and use them to estimate values of the parameters in Equation E-11:

• The background rate is so low that we expect to see no tumors in the control group in most experiments.

• The single-exposure doses of each agent are chosen to yield identical responses in the bioassay. This condition may be expressed within the context of the model depicted in Equation E-11 as $B_1x_1 = B_2x_2$, which, for ease of notation, we set equal to the constant V.

• The doses of the single agents are chosen to be as low as possible such that an observed result that would be equal to the expected value of the response would indicate a statistically significant increase in tumor rates over the controls. That is done so that a response can be measured for each agent, but the greatest possible leeway is left for measuring the joint effect of two agents. Assuming zero tumors in the control group, because the background rate is low, five or more tumors are needed in a test group for the response to be considered statistically significant at the 0.05 level, according to a Fisher exact 2×2 test. To meet that condition, we select our exposure dose so that the probability that the tumor increase is statistically significant is 0.5. That implies that the probability of obtaining four or fewer tumors in a single-agent test group is also 0.50. The probability that a single animal will develop a tumor at the single-agent dose under the conditions of the experiment is obtained from Equation E-11 by setting $B_1x_1 = 0$ and $B_2x_2 = V$, or vice versa. This yields a response expressed in the form

$$P = 1 - \exp\{-A(1 + V)\}. \tag{E-12}$$

The probability of four or fewer tumors is obtained from a binomial distribution with $n = 50$ and P defined as in Equation E-12. The probability is set equal to 0.5 so that the value of $A(1 + V)$ can be estimated. The relationship may be expressed as

$$\sum_{s=0}^{4} \binom{50}{s} [1 - e^{-A(1 + V)}]^{s} e^{-A(1 + V)(50 - s)} = 0.5, \tag{E-13}$$

which is solved numerically to obtain the solution,

$$A(1 + V) = 0.097282. \tag{E-14}$$

To estimate values of the individual terms A and V, we must impose an additional condition on the bioassay:

• Interaction cannot be measured quantitatively if the joint response is 100%. To obtain the maximal measurable interactive effect, we assume that the joint interactive effect will yield a 100% response 50% of the time, so a large interaction effect would be measurable 50% of the time. The probability that all 50 animals will respond at the joint exposure dose is obtained from the binomial distribution with $P = 1 - \exp\{-A(1 + V)^2\}$ and $n = 50$, where P is obtained from Equation E-11 when $B_1x_1 = B_2x_2 = V$. The relationship may be expressed as

$$[1 - e^{-A(1 + V)^2}]^{50} = 0.5, \qquad \text{(E-15)}$$

which has the solution

$$A(1 + V)^2 = -\ln[1 - (0.5)^{1/50}] = 4.28546. \qquad \text{(E-16)}$$

The numerical values for A and V can be obtained by solving Equations E-14 and E-16 simultaneously. Dividing Equation E-16 by Equation E-14 gives the result

$$\frac{A(1 + V)^2}{A(1 + V)} = 1 + V = \frac{4.28546}{0.097282} = 44.05187.$$

$$V = 44.05187 - 1 = 43.05187,$$

and

$$A = 0.097282/44.05187 = 0.0022084.$$

To obtain estimates of values of the response-model parameters B_1 and B_2, the forms with which Equation E-11 is expressed, one additional condition is needed: exposure doses x_1 and x_2, which would give the single-agent responses, are required. They represent a scaling factor to redefine the equation for specific exposure potencies. To obtain a specific equation, we arbitrarily assume that the low-dose response to each of the single agents resulted from exposures at $x_1 = 0.1$ and $x_2 = 0.2$; agent 1 is twice as potent as agent 2. Because $B_1x_1 = B_2x_2 = V$, it follows that

$$B_1 = 43.05187/0.1 = 430.5187,$$

and

$$B_2 = 43.05187/0.2 = 215.2593.$$

Substituting the numerical estimates of A, B_1, and B_2 into Equation E-11 yields the following dose-response model:

$$P(x_1,x_2) = 1 - \exp\{-0.002208(1 + 430.5187x_1)(1 + 215.2593x_2)\}, \qquad \text{(E-17)}$$

or

$$P(x_1,x_2) = 1 - \exp\{-[0.002208 + 0.9507x_1 + 0.4754x_2 + 204.6552x_1x_2]\}. \qquad \text{(E-18)}$$

Given this model, it is possible to estimate the risk at any exposures x_1 and x_2. The model derived here predicts an interaction effect that is one of the largest that can be estimated from a 2 × 2 balanced-design bioassay with 50 or fewer animals per group. If designs as assumed here were used to estimate joint effects, interactions of the magnitude assumed here would be among the largest observed.*

ESTIMATION OF JOINT EFFECTS OF ENVIRONMENTAL EXPOSURE WITH PREVIOUSLY DERIVED NUMERICAL MODEL

The model depicted in Equation E-18 will be considered to be the true underlying dose-response relationship. That model has a large interaction-synergistic effect built into it; it is about the largest that could be measured with a classic 2 × 2 balanced design with 50 animals per exposure group and that would allow the statistical estimation of the multistage model.

In many situations, the results of multiple-agent exposure experiments are not available, so estimates of cancer risk associated with environmental exposures must be based on the results of single-agent experiments.

In this section, we evaluate the error that the assumption of low-dose additivity could introduce. For the purpose of the evaluation, the true underlying dose-response relationship for a single agent is assumed to be known and to be the marginals from Equation E-18. The term "marginals" refers to the response obtained from exposure to one agent when there is no exposure to the second agent. Thus, the excess cancer risk associated with exposure to the first or second agent is

$$P^*(x_1) = P(x_1,0) - P(0,0)$$

or (E-19)

$$P^*(x_2) = P(0,x_2) - P(0,0),$$

respectively, with $P(x_1,x_2)$ as defined in Equation E-18. Assuming additivity, the estimated joint excess risk associated with both exposures is

$$\hat{P}^*(x_1,x_2) = P^*(x_1) + P^*(x_2). \quad \text{(E-20)}$$

The error introduced by that assumption can thus be evaluated by comparing $\hat{P}^*(x_1,x_2)$ with the true excess risk of

$$\hat{P}^*(x_1,x_2) = P(x_1,x_2) - P(0,0). \quad \text{(E-21)}$$

*To our knowledge, an interaction of the magnitude of this example has never been observed; however, an infinite interaction is theoretically consistent with theory underlying the multistage model. A bioassay design similar to that defined here would not, however, allow the estimation of a multistage model with more extreme interaction than that given here.

The true increased risk associated with exposure to one agent conditional on exposure to a second agent may be written as:

$$P^*(x_1|x_2) = P(x_1, x_2) - P(0, x_2),$$

and (E-22)

$$P^*(x_2|x_1) = P(x_1, x_2) - P(x_1, 0).$$

We can evaluate the error introduced by the approximations for various values of x_1 and x_2. The results of this analysis are shown in Table E-1. From Table E-1, the following trends are noted:

- The synergistic effect is extreme at high doses (i.e., $x_1 = 0.1$, $x_2 = 0.2$). Each agent by itself gives about a 10% response, but jointly they yield a 99% response. In this case, the additivity assumption clearly does not hold.
- If one agent remains at a high dose (i.e., $x_1 = 0.1$) and the other is reduced by 2 or more orders of magnitude, the additivity approximation is reasonably good. The same is true for the conditional risk for the high-exposure agent. However, the conditional risk for the low-exposure agent is 40 times higher than predicted by the additivity assumption at smaller exposures.
- If exposure to both agents is reduced by 2 orders of magnitude, the additivity assumption is reasonably good. If exposure to both agents is reduced by 4 orders of magnitude, there is virtually no difference between the true risk and that predicted by the additivity assumption.

PRACTICAL IMPLICATIONS DISCERNIBLE FROM NUMERICAL INVESTIGATION OF GENERALIZED MULTISTAGE MODEL

If the previous examples were representative of the types of dose-response relationships encountered in practice, the following conclusions can be drawn:

- Agents with high environmental exposure—such as background radiation, cigarette smoke, and some workplace exposures—must be investigated very carefully, because, if they act on the same cell type in the same organ, the potential for a strong synergistic effect is great.
- The excess cancer risk at low doses from an agent that acts on the same cell type in the same organ as another agent(s) to which exposure at high levels occurs (e.g., cigarette smoke) could be seriously underestimated in an animal bioassay, because the bioassay ignores the effects of the other agent(s), such as cigarette smoke, on the estimated augmented risk.
- When all environmental exposures are 3–4, or more, orders of magnitude below that associated with observable effects in bioassays or in epidemiology studies, additivity assumptions can provide a reasonable approximation of the joint risk.

TABLE E-1 Evaluation of the Error Introduced by the Use of the Simple Additivity and Independence Assumptions

Cancer Risk Associated with Exposure to: Agent 1 x_1	Agent 2 x_2	True Joint Risk $P^*(x_1,x_2)$	Predicted Joint Risk Assuming Additivity $P^*(x_1,x_2)$	$\frac{P^*(x_1,x_2)}{P^*(x_1,x_2)}$	True Augmented Risk $P^*(x_1 \vert x_2)$	Predicted Augmented Risk $P^*(x_1)$	$\frac{P^*(x_1 \vert x_2)}{P^*(x_1)}$	True Augmented Risk $P^*(x_2 \vert x_1)$	Predicted Augmented Risk $P^*(x_2)$	$\frac{P^*(x_2 \vert x_1)}{P^*(x_2)}$
1×10^{-1}	2×10^{-1}	9.84×10^{-1}	1.81×10^{-1}	5.436	8.94×10^{-1}	9.05×10^{-2}	−9.878	8.94×10^{-1}	9.05×10^{-2}	−9.878
1×10^{-1}	2×10^{-2}	3.34×10^{-1}	9.99×10^{-2}	3.343	3.27×10^{-1}	9.05×10^{-2}	−3.613	2.43×10^{-1}	9.44×10^{-3}	−25.742
1×10^{-1}	2×10^{-3}	1.28×10^{-1}	9.14×10^{-2}	1.400	1.27×10^{-1}	9.05×10^{-2}	1.403	3.75×10^{-2}	9.48×10^{-4}	39.557
1×10^{-1}	2×10^{-4}	9.43×10^{-2}	9.06×10^{-2}	1.041	9.42×10^{-2}	9.05×10^{-2}	1.041	3.79×10^{-3}	9.51×10^{-5}	39.853
1×10^{-1}	2×10^{-5}	9.09×10^{-2}	9.05×10^{-2}	1.004	9.09×10^{-2}	9.05×10^{-2}	1.004	3.79×10^{-4}	9.51×10^{-6}	39.853
1×10^{-3}	2×10^{-3}	2.31×10^{-3}	1.90×10^{-3}	1.216	1.36×10^{-3}	9.51×10^{-4}	1.430	1.36×10^{-3}	9.51×10^{-4}	1.430
1×10^{-5}	2×10^{-5}	1.90×10^{-5}	1.90×10^{-5}	1.000	9.51×10^{-6}	9.51×10^{-6}	1.000	9.51×10^{-6}	9.51×10^{-6}	1.000

- Because of the previous result, it might be counterproductive to test the agents in a laboratory experiment, because interaction observed at high exposures could vanish at low exposures. In such cases, little additional information about low-dose behavior under the assumed model would be gained from joint-exposure experiments, compared with that from single-agent experiments.

The next section presents a more general mathematical treatment of the conditions under which the low-dose additivity assumption might be expected to be accurate.

EFFECT OF BACKGROUND TUMOR RATE ON ADDITIVITY ASSUMPTION

Background tumor rate has an effect on the accuracy of the additivity assumption. For the multistage model, the larger the background rate, the faster the dose-response relationships approach linearity (Hoel, 1980). As a result, the additivity assumption is more accurate for higher background rates at levels of exposure that yield equivalent responses in the laboratory. The example discussed in the previous section had a background rate of only 0.2%, which would tend to give a less accurate approximation of risk than examples with higher background rates. To demonstrate the effect of background rate on the additivity assumption, consider the following example. Assume that agent 1 at exposure x_1 and agent 2 at exposure x_2 produce equal effects on transition rates between stages and that background transition rates are equal for each stage. These conditions yield a model that changes rapidly with increasing dose, so the deviation from additivity is at the maximum for all models of this class. The model for which these conditions hold may be expressed as

$$P(x_1, x_2) = 1 - \exp\{-A(1 + Sx_1)(1 + Sx_2)\} = 1 - \exp\{-A(1 + Sx)^2\},$$

where $x_1 = x_2 = x$, where the background tumor rate is $1 - e^{-A}$.

For all cases, risk at $x = 1$ is set equal to 0.5. The extent of deviation of the true cancer risk from that predicted with the assumption of additivity is shown in Table E-2 for doses of $x = 0.01$ or 0.001 and various assumed background rates.

Note that the accuracy of the additivity approximation decreases as the background rate decreases. However, for an exposure that is 3 orders of magnitude lower than that which yields a response of 0.5, the accuracy is very good for the full span of background rates.

ADDITIVITY OF EXCESS RISKS AT LOW DOSES

Interactive effects at moderate to high doses have been demonstrated in toxicologic experiments and observed in epidemiologic studies, but existence of

TABLE E-2 Effect of Background Rate on Accuracy of Additivity Approximation Under the Assumption of the Multistage Model Defined in Equation E-11

True Background Rate	Exposure Level for Both Agents	Deviation of Additivity Approximation from True Risk after Subtracting Background, %
0.1	0.01	−0.73
	0.001	−0.10
0.01	0.01	−3.39
	0.001	−0.34
0.001	0.01	−11.23
	0.001	−1.18

such effects at low doses is not amenable to direct investigation, because of the low response rates generally associated with low doses. This section identifies conditions under which the additivity assumption can be expected to hold at sufficiently low doses.

MULTISTAGE MODEL

Consider first the multistage model in the case of joint exposure to two chemicals C_1 and C_2 given in Equation E-8. The probability that a tumor will occur by time t after exposure to C_1 and C_2 at doses of x_1 and x_2, respectively, is given by:

$$P(x_1,x_2,t) = 1 - \exp\left\{-N\frac{t^k}{k!}\prod_{i=0}^{k-1}\lambda_i\right\}$$

$$\simeq Nt^k/k!\left(\prod_{i=0}^{k-1}\lambda_i\right), \qquad \text{(E-23)}$$

for small values of x_1 and x_2, where

$$\lambda_i = \alpha_i + \sum_{j=1}^{2}\beta_{ij}x_j \qquad \text{(E-24)}$$

denotes the transition intensity function for the ith stage ($i = 0,1, \ldots, k-1$). At any fixed time t, it follows from Equation E-23 that the excess risk at low doses may be written as

$$\Pi(x_1,x_2) = P(x_1,x_2) - P(0,0)$$

$$\simeq c_1x_1 + c_2x_2, \qquad \text{(E-25)}$$

where

$$c_j = (Nt^k)/k! \sum_{i=0}^{k-1} (\prod_{i' \neq i} \alpha_{i'})\beta_{ij} > 0. \qquad \text{(E-26)}$$

Because the excess risk for chemical C_j is approximately $c_j x_j$ for sufficiently small x_i, it follows from Equation E-25 that the excess risks for C_1 and C_2 are additive at these doses with the multistage model in Equation E-23. Note that the same result also demonstrates that the excess risk associated with each chemical will be nearly linear in dose at these doses.

ADDITIVE-BACKGROUND MODELS

For exposure to single chemicals, Crump et al. (1976) have established other conditions under which low-dose linearity will obtain. The same argument may be invoked in the case of joint exposure to two substances C_1 and C_2, to demonstrate low-dose linearity and hence additivity of the excess risks associated with C_1 and C_2 separately. Specifically, suppose that the joint response rate can be expressed as

$$P(x_1, x_2) = H(x_1 + \delta_1, x_2 + \delta_2), \qquad \text{(E-27)}$$

where H is an increasing function of both x_1 and x_2 with both first partial derivatives D_1H and D_2H strictly positive. Here, $\delta_1 > 0$ and $\delta_2 > 0$ denote the effective background doses of x_1 and x_2, respectively, leading to a spontaneous-response rate $P(0,0) = H(\delta_1,\delta_2)$. A Taylor expansion of $P(x_1,x_2)$ shows that, for sufficiently small x_1 and x_2,

$$\Pi(x_1, x_2) \simeq c_1x_1 + c_2x_2, \qquad \text{(E-28)}$$

where $c_i = D_iH\,(\delta_1,\delta_2) > 0$. The model in Equation E-27 has been termed an additive-background model, because the doses x_1 and x_2 of C_1 and C_2 are considered to act in an additive manner in conjunction with the effective background doses δ_1 and δ_2.

MIXED-BACKGROUND MODELS

For single chemicals, the result of Crump et al. (1976) was further generalized by Hoel (1980) to include combinations of both additive and independent background. In the present context, combinations of independent and additive background may be represented by the model

$$P(x_1, x_2) = \gamma + (1 - \gamma)H(x_1 + \delta_1, x_2 + \delta_2), \qquad \text{(E-29)}$$

where $0 \leq \gamma < 1$ denotes the probability of a spontaneous tumor's occurring independently of those induced by C_1 or C_2 or the effective background doses δ_1 and δ_2. The same argument as used for the additive-background model in Equation E-27 may be used to show, for sufficiently small x_1 and x_2, that

$$\Pi(x_1,x_2) \simeq c_1^* x_1 + c_2^* x_2 \quad \text{(E-30)}$$

under the mixed-background model in Equation E-29, with $c_j^* = (1 - \gamma)c_j$ for $j = 1, 2$.

Multiplicative-Risk Models

In some cases, the interaction between two substances might be well described by a multiplicative model in which the relative risk associated with the mixture is the product of the relative risks associated with the components. Let $p_{10} = P(x_1, 0)$ and $p_{01} = P(0, x_2)$ denote the respective probabilities of tumor occurrence when chemicals C_1 and C_2 are administered separately, and let $p_{00} = P(0,0)$ denote the spontaneous-response rate. The corresponding relative risks associated with chemicals C_1 and C_2 are thus given by $r_{10} = p_{10}/p_{00}$ and $r_{01} = p_{01}/p_{00}$, respectively. Letting $p_{11} = P(x_1,x_2)$, the relative risk $r_{11} = p_{11}/p_{00}$ when both C_1 and C_2 are administered simultaneously is given by

$$r_{11} = r_{10} r_{01} \quad \text{(E-31)}$$

with the multiplicative-risk model (Siemiatycki and Thomas, 1981).

With this model, the excess risks associated with C_1 and C_2 will be approximately additive when both r_{10} and r_{01} are sufficiently close to 1. To see this, note that, from Equation E-31,

$$r_{11} - 1 \simeq (r_{10} - 1) + (r_{01} - 1) \quad \text{(E-32)}$$

because $\log r \simeq r - 1$ for r near unity. Thus,

$$p_{11} - p_{00} \simeq (p_{10} - p_{00}) + (p_{01} - p_{00}), \quad \text{(E-33)}$$

when the relative risks r_{10} and r_{01} are close to 1. This implies that, for two substances whose interaction follows a multiplicative-risk model, as with cancer of the oral cavity induced by smoking and alcohol consumption (Rothman and Keller, 1972; Tuyns et al., 1977), the excess risks will be effectively additive at doses where the relative risks are sufficiently small (e.g., less than 1.05).

PREDICTION OF RISK AT LOW DOSES

To predict the potential carcinogenic effects of joint exposure to two or more substances at low doses, it is necessary to extrapolate downward from higher doses that induce measurable rates of response (Brown, 1984; Clayson and

Krewski, 1986). When data on the effects of joint exposures are available, that may be done by modeling tumor-occurrence rates as a function of x_1 and x_2 of chemicals C_1 and C_2. The response surface described by the fitted model may then be used to extrapolate to lower doses.

The preceding section identified models under which the excess risks associated with simultaneous exposures to C_1 and C_2 would be additive at sufficiently low doses. When those conditions are satisfied, it is sufficient to predict the excess risks associated with C_1 and C_2 separately, because the excess risk associated with joint exposure to the two substances is well approximated by the sum of the excess risks associated with exposures to C_1 and C_2 alone. That may be done with separate dose-response curves for C_1 and C_2, so that the prediction of low-dose risks does not require mapping the response surface associated with joint exposure to the two substances in this case. When joint-exposure data are available, they may be used not only to identify interactions that can exist at moderate to high doses, but also to assess the extent to which such interactions can be reduced at lower doses. To make full use of joint-exposure data in predicting risks at low doses, we require a model for the probability of tumor induction $P(x_1, x_2)$ by time t after exposure to chemicals C_1 and C_2 at constant doses x_1 and x_2. For example, the multistage model in Equation E-11 is of the form

$$\begin{aligned} P(x_1, x_2) &= 1 - \exp\{-A(1 + B_1x_1)(1 + B_2x_2)\} \\ &= 1 - \exp\{-(\Theta_0 + \Theta_1x_1 + \Theta_2x_2 + \Theta_3x_1x_2)\}, \end{aligned} \qquad \text{(E-34)}$$

where the Θ_is are subject to the nonlinear constraint $\Theta_0\Theta_3 = \Theta_1\Theta_2$, as well as the linear constraints $\Theta_i \geq 0$ $(i = 0,1,2,3)$.

To avoid nonlinear constraints in fitting models of the form of Equation E-34, the linear constraints $\Theta_i \geq 0$ may be invoked, as has been done in fitting the multistage model to data involving exposure to only a single substance (Krewski and Van Ryzin, 1981; Armitage, 1982). The linear constraints represent a broader class of models than the nonlinear constraints arising in the derivation of the model, but greatly simplify the statistical problems involved in fitting the models to experimental data. (Note that, although the linear constraints admit the possibility that Θ_1 or Θ_2 is zero, upper confidence limits on values of these parameters will necessarily be greater than zero, so low-dose linearity is preserved.)

More generally, we consider the model

$$P(x_1, x_2) = 1 - \exp\left\{-\left(\Theta_0 + \sum_{i=1}^{k}(\Theta_1^i x_1^i + \Theta_2^i x_2^i) + \sum_{i=1}^{k}\sum_{j=1}^{k}\Theta_3^{i,j}x_1^i x_2^j\right)\right\}, \qquad \text{(E-35)}$$

where the values of parameters are constrained to be nonnegative. To fit this model to data on a response surface $P(x_1, x_2)$ that is highly curvilinear, it might be necessary to use a value of k in excess of 1 or 2. The model involves $(k + 1)^2$ unknown parameters, so experiments with at least this many exposure groups are necessary. With $k = 3$, for example, at least 16 treatment groups would be required.

The process of describing the response surface would be simpler, if a sufficiently flexible model involving fewer parameters were used. For example, the four-parameter logistic model

$$P(x_1, x_2) = [1 + \exp\{-(\Theta_0 + \Theta_1 x_1 + \Theta_2 x_2 + \Theta_3 x_1 x_2)\}]^{-1} \quad \text{(E-36)}$$

might be used to describe the response surface $P(x_1, x_2)$ within the observable-response range. In analogy with the simple linear extrapolation procedure proposed by Van Ryzin (1980), excess risks at low doses could then be predicted by extrapolating along straight lines joining the response surface at points $P(x_1^*, x_2^*)$ and the origin $P(0,0)$, where (x_1^*, x_2^*) is chosen so that the excess risk $\pi(x_1^*, x_2^*) = \pi^*$, where π^* is chosen to be as small as possible, yet still within the experimentally measurable response range (e.g., $0.01 \leq \pi^* \leq 0.10$). The points $P(x_1^*, x_2^*)$ may be estimated by fitting a model, such as that in Equation E-36, to experimental data involving joint exposures to C_1 and C_2.

SUMMARY

Empirical tests of joint action in laboratory settings, which typically are based on high doses, provide little insight into the corresponding effects at low environmental doses. The information available on the mechanisms of tumor induction by chemicals strongly suggests that their relationship with dose is not linear. Quantitative models of the carcinogenic process, such as the multistage model of Armitage and Doll, also reflect such nonlinearity in the response to mixtures. Although a multiplicative exposure effect sometimes dominates at high doses, further exploration of this model indicates that the joint effect will be additive (that is, close to the sum of the individual effects) at sufficiently low doses. A newer model, that of Moolgavkar and Knudson (1981), is more biologically specific and yields essentially the same conclusion. (The conclusion depends on the assumption that the augmented risks of the chemical are small, with respect to the natural background rate of the tumor.) Additivity at low doses was also demonstrated under a general class of additive background models and under the multiplicative risk model when the relative risk for each component in the mixture is small.

In addition, it can be shown for a broad range of mathematical dose-response models that the joint risk associated with a complex mixture can be determined on the basis of the background risk and the risks associated with the individual components. The models for which that is true include many widely used dose-response models, such as the linear, multiplicative, logistic, and multistage models. The results suggest the utility and desirability of identifying appropriate models.

Prediction of low-dose risks generally requires the extrapolation of results obtained at higher doses that induce measurable tumor-response rates. When the excess risks associated with exposures to mixture components can be reasonably considered to be additive, that can be done by downward extrapolation of the dose-response curves for the individual components. Otherwise, it will be necessary to model the tumor-response surface associated with the joint exposure to two or more carcinogens. In general, the use of joint-exposure data is to be preferred, in that it provides some empirical evidence on the magnitude of any interactions that might be present even at lower doses.

REFERENCES

Armitage, P. 1982. The assessment of low-dose carcinogenicity. Biometrics 28(Suppl.):119–129.

Armitage, P., and R. Doll. 1954. The age distribution of cancer and a multi-stage theory of carcinogenesis. Br. J. Cancer 8:1–12.

Berenblum, I., and P.A. Shubik. 1949. An experimental study of the initiating stage of carcinogenesis, and a re-examination of the somatic cell mutation theory of cancer. Br. J. Cancer 3:100–118.

Brown, C. C. 1984. High- to low-dose extrapolation in animals, pp. 57–79. In J. V. Rodricks and R. G. Tardiff, Eds. Assessment and Management of Chemical Risks. ACS Symposium Series 239. American Chemical Society, Washington, D.C.

Clayson, D. B., and D. Krewski. 1986. The concept of negativity in experimental carcinogenesis. Mutat. Res. 167:233–240.

Crump, K. S., D. G. Hoel, C. H. Langley, and R. Peto. 1976. Fundamental carcinogenic processes and their implications for low dose risk assessment. Cancer Res. 36:2973–2979.

Hoel, D. G. 1980. Incorporation of background in dose-response models. Fed. Proc. 39:73–75.

Krewski, D., and J. Van Ryzin. 1981. Dose response models for quantal response toxicity data, pp. 201–231. In M. Csörgö, D. A. Dawson, J. N. K. Rao, and A. K. Md. E. Saleh, Eds. Statistics and Related Topics. North-Holland, New York.

Moolgavkar, S. H., and A. G. Knudson, Jr. 1981. Mutation and cancer: A model for human carcinogenesis. J. Natl. Cancer Inst. 66:1037–1052.

Rothman, K., and A. Keller. 1972. The effect of joint exposure to alcohol and tobacco on risk of cancer of the mouth and pharynx. J. Chron. Dis. 25:711–716.

Siemiatycki, J., and D. C. Thomas. 1981. Biological models and statistical interactions: An example from multistage carcinogenesis. Intl. J. Epidemiol. 10:383–387.

Tuyns, A. J., G. Péquignot, and O. M. Jensen. 1977. Oesophagal cancer in Ille-et-Vilaine in relation to alcohol and tobacco consumption: Multiplicative risks. Bull. Cancer 64:45–60. (In French; English summary.)

Van Ryzin, J. 1980. Quantitative risk assessment. J. Occup Med. 22:321–326.

Whittemore, A., and J. B. Keller. 1978. Quantitative theories of carcinogenesis. SIAM Rev. 20:1–30.

F

Developmental Toxicology

Many hazardous environmental chemicals seem to have toxic effects in the offspring of the persons exposed, including birth defects and spontaneous abortion. Therefore, in assessing the risks associated with chemical mixtures, one should investigate their possible teratologic effects. As a step in assessing these risks, animal bioassays are performed at increasing doses of the chemicals to be studied. The end points of interest in the bioassays are fetal death, failure to grow, structural and functional abnormalities, and behavioral deficiencies. Other elements of the reproductive cycle are not considered.

A typical animal bioassay is carried out by exposing a female before or during pregnancy or a male before mating, observing their offspring (if any), and studying the end points in relation to the doses administered. The toxic responses are recorded in the offspring, but the experimental units are the females. Mantel (1969) recognized that an inherent characteristic of this type of data is the so-called litter effect, which is the tendency for littermates to respond more similarly than animals from different litters. As a result, the quantal dose-response information obtained from teratologic experiments is different from the usual dose-response data, based on effects (tumors, etc.) observed in directly exposed animals. The methods of analyzing such data with dose-response models must take the difference into account. We review these methods briefly here.

STATISTICAL METHODS FOR DEVELOPMENTAL-TOXICOLOGY DATA

In the typical developmental-toxicology experiment with animals, the data collected are expressed as in Table F-1.

TABLE F-1 Characteristics of Data from Developmental-Toxicology Experiments with Animals

		Test Groups	
	Control Groups	Low	High
Dose	$d_0 = 0$	d_1	d_m
No. females	n_0	n_1	n_m
No. offspring	$s_{01}, \ldots, s_{0n_0}$	$s_{11}, \ldots, s_{1n_1}$	$s_{ml}, \ldots, s_{mn_m}$
No. offspring with defect	$x_{01}, \ldots, x_{0n_0}$	$x_{11}, \ldots, x_{1n}$	$x_{ml}, \ldots, s_{mn_l}$

As Table F-1 suggests, the experiment is usually carried out at m increasing doses $d_1 < \ldots < d_m$ in the control group, at which there is no exposure, $d_0 = 0$. The experimenter randomizes n_i females to the ith group and exposes them to dose d_i. For each female in dose group d_i, the experimenter records the litter size s_{ij}, with $j = 1, 2, \ldots, n_i$, and x_{ij} the number of offspring with a specified toxic defect. Let p_{ij} be the probability that an offspring in the ij^{th} litter will be defective. In analyzing such quantal response data statistically, the probability p_{ij} is usually assumed to vary from female to female. Within a litter, the number of defects, x_{ij}, is assumed to have a binomial distribution with defect rate p_{ij} for the s_{ij} offspring. Such modeling allows for estimating within-litter and between-litter variation.

Haseman and Kupper (1979) and Van Ryzin (1985a) have discussed a number of statistical procedures for such a model. In particular, one can estimate the mean defect rate at each dose (Van Ryzin, 1975), test whether a particular dose group differs from the controls in defect rate (Hoel, 1974), and test whether the mean defect rate increases with dose by using Jonckheere's nonparametric trend test (see Lehman, 1975, p. 233). A parametric approach was proposed by Williams (1975), who fitted the beta-binomial model to teratologic data. In still another approach, Kupper and Haseman (1978) presented a correlated binomial model, which, in contrast with the beta-binomial model, allows for negative as well as positive intralitter correlation. In both previous cases, the authors estimate the relevant parameters with maximum-likelihood methods. A likelihood ratio test is used to compare results in the treatment and control groups. (For further discussion of the procedures, see Kupper and Haseman, 1978; Van Ryzin, 1985a.)

The statistical procedures described make it possible to test for the presence or absence of a dose-response relationship in developmental-toxicology bioassays for various developmental end points or for fetal toxicity. It is also possible to assess the effects of dose on conceptus resorption, which is thought to be the rough equivalent in animal studies of spontaneous abortion in humans. That can be done by standard analysis methods in which s_{ij} is defined as the number of implants of the jth female in dose group d_i and x_{ij} is the number of resorbed fetuses.

As is the case for directly observable quantal response data, such as carcinogenicity data, the establishment of a dose-response relationship raises the question of how to estimate the risks associated with low environmental exposure to a substance, given the responses observed at (usually) higher experimental doses. That involves low-dose extrapolation through the use of some dose-response model. The models for developmental-toxicology data proposed in the literature are discussed in the next section, which also includes a new test for detecting a response trend with dose for teratology data.

DOSE-RESPONSE MODELS FOR DEVELOPMENTAL EFFECTS

The literature on development of dose-response models for developmental effects or fetal death is rather sparse. A broad set of models must be considered, because no process to explain the effects is generally accepted, and many end points are involved. It is commonly held that developmental effects and fetal toxicity have a threshold dose-response relationship and that safety factors can be used to protect public health. However, the applicability of threshold models for all developmental end points is open to debate. For example, Jusko (1972, 1973) examined two classes of developmental toxicants—those that appear to have a threshold and those that do not. In particular, Jusko's analyses (1972, 1973) would indicate that thalidomide and cyclophosphamide are non-threshold developmental toxicants.

At least two attempts to model dose-response relationships for developmental effects have been reported. Jusko (1972) has argued for a drug receptor model for fetal exposure based on pharmacokinetics. Assuming a "hit theory," he modeled the fraction of intact or normal fetuses as an exponentially decreasing function of dose for some toxicants. In the notation used here, s_{ij} is the number of implants at dose d_i in female j, and x_{ij} is the number of resorbed, dead, or defective fetuses from this female. Thus, the response rate (live births), $(s_{ij} - x_{ij})/s_{ij}$, of normal fetuses would have a mean or expected value of the form

$$\text{mean}\,[(s_{ij} - x_{ij})/s_{ij}] = \exp(-Kd_i), \tag{F-1}$$

for $K > 0, j = 1, \ldots, n_i$, and $i = 0, 1, 2, \ldots, m$. Equation F-1 assumes that there is no background response rate, but if there were one, it could be included by substituting $\exp(-Kd_j + d_0)$ in the right side of the equation. For the no-background-rate case, Equation F-1 can be rewritten as

$$\text{mean}\,(1 - P_{ij}) = \exp(-Kd_i), \tag{F-2}$$

where P_{ij} is the proportion of dead, resorbed, or defective fetuses in the offspring of the jth female exposed to dose d_i.

Jusko (1972) argued further that other developmental toxicants that must affect r receptor sites ($r > 1$) before having a toxic effect in offspring would follow Equation F-3:

$$\text{mean}\,[(s_{ij} - x_{ij})/s_{ij}] = 1 - [1 - \exp(-Kd_i)]^r, \quad \text{(F-3)}$$

where $r > 1$ and no background effect is assumed. Equations F-1 and F-3 may be rewritten as one equation in terms of the mean defect rate at dose d_i:

$$\mu(d_i) = \text{mean}\,(x_{ij}/s_{ij}) = [1 - \exp(-Kd_i)]^r, \quad \text{(F-4)}$$

for $r \geq 1$. A model like Equation F-4 could be used for low-dose extrapolation by extending methods of low-dose extrapolation now in existence for directly observable quantal dose-response models (Krewski and Van Ryzin, 1981). However, such methods have not yet been fully reported. When case $r = 1$, Equation F-4 implies that the dose-response is linear at low doses (low responses); that is, for d near zero, Equation F-4 becomes

$$\mu(d) = 1 - \exp(-Kd) \simeq Kd. \quad \text{(F-5)}$$

Equation F-5 implies that at times linear extrapolation at low doses is appropriate. Assuming that the above model is valid, one could combine the low-dose effects of simple mixtures in a manner similar to that suggested for carcinogenesis models (see Appendix E).

A second attempt at low-dose extrapolation with dose-response models for developmental toxicology data has been proposed by Rai and Van Ryzin (1985). They modeled the variation of the defect rate or proportion P_{ij} by taking

$$P_{ij} = \lambda(d_i) \exp[-(\beta_1 w_{1j} + \ldots + \beta_k w_{kj})], \quad \text{(F-6)}$$

where $\beta_1 w_{1j} + \ldots + \beta_k w_{kj} \geq 0$. Here, $\lambda(d_i)$ denotes a baseline defect rate or proportion at dose d_i, and the exponential term represents the modification of that baseline rate due to the particular covariates or mitigating factors $w_{\nu j}$, $\nu = 1, \ldots, k$, for the jth female. For example, one could include such characteristics of the dam as maternal weight loss and litter size as covariates in this model. However, the covariates, $w_{\nu j}$, need to be observable. The baseline response $\lambda(d)$ in Equation F-6 also needs to be specified. Rai and Van Ryzin (1985) studied a particular case of Equation F-6 by taking

$$\lambda(d) = 1 - \exp[-(\alpha + \beta d)] \quad \text{(F-7)}$$

for $\alpha > 0$ and $\beta > 0$ and including a single covariate, the litter size (s_{ij}). If Equation F-7 holds, the female baseline defect-response rate follows a one-hit model. In this case, low-dose extrapolation based on $\lambda(d)$ would be linear. Extensions of the work of Rai and Van Ryzin could be developed for low-dose extrapolation based on any specified $\lambda(d)$ (Van Ryzin, 1985a). If Equation F-7 is appropriate and low-dose extrapolation for each chemical is based on $\lambda(d)$, corresponding low-dose extrapolation for mixtures would be dose-additive, as is the case for the multistage model in carcinogenic studies (see Appendix E).

A related problem was addressed by Munera (1986), who proposed a trend test for teratologic data. In this approach, the random response probability, P_{ij}, is allowed to vary within the dose group d_i according to an unknown distribution, G_i. However, the mean of the distribution is assumed to follow a specified regression function that might depend on dose. For example, if Equation F-2 holds, one could take

$$\text{mean}\ (P_{ij}) = 1 - e^{(-Kd_i)}.$$

The statistical trend test derived from this model is based on empirical Bayesian statistical procedures applied to the unknown distribution G_i and tests whether the regression function increases with dose.

LIMITATIONS AND ADVANTAGES OF PROCEDURES THAT USE DEVELOPMENTAL END POINTS

The statistical procedures for analyzing developmental-toxicity studies discussed previously try to account for the greater than binomial variability found in the data collected from such experiments. However, no model seems to have a clear advantage over the others (see Haseman and Kupper, 1979). Little effort has been made to compare the advantages of the various approaches systematically.

The trend test proposed by Munera (1986) is also new. An alternative way of estimating the regression parameters might be more efficient than their empirical Bayesian estimates. This method merits further investigation.

Similarly, the main problem in using the dose-response models mentioned in their article is that they are not yet fully developed and have not yet received general scientific acceptance. That is, the dose-response models proposed for developmental data are not yet clearly justified by the biologic evidence. The models discussed above appear to fit some data well, but it is possible that no single dose-response model for developmental toxicity will be generally acceptable.

The results to date do suggest that several developmental toxicants fall into the class of materials exhibiting low-dose linearity. If the parallel to carcinogenesis holds—where nonlinearities in the observed dose range due to nonlinear kinetics are consistent with low-dose linearity (Hoel et al., 1983; Rai and Van Ryzin, 1984; Van Ryzin, 1985b)—then low-dose linear extrapolation would be appropriate in assessing the risk of developmental effects.

Data on developmental toxicity of mixtures and their components can be gathered more quickly than data on carcinogens, inasmuch as animal reproductive-developmental bioassays can be carried out relatively rapidly (about 2 months), as opposed to carcinogenicity assays (approximately 2 years). Thus, the possibility of economical testing of a variety of mixtures for developmental toxicity is greater than the possibility of economical testing for carcinogenicity.

SUMMARY

Developmental-toxicity end points present an opportunity for assessing the joint actions of chemicals. In the usual laboratory experiment, each exposed female provides several offspring, so both litter size and incidence of abnormalities within litters provide an index of toxicity. In addition, the incidence of resorptions can be assessed. Dose-response functions can be extracted from the various end points, but the crucial problem is low-dose extrapolation. Models that have been devised for that purpose suggest that low-dose extrapolation for mixtures would likely be dose-additive, but there is no general acceptance of these models.

REFERENCES

Haseman, J. K., and L. L. Kupper. 1979. Analysis of dichotomous response data from certain toxicological experiments. Biometrics 35:281–293.

Hoel, D. G. 1974. Some statistical aspects for experiments for determining the teratogenic effects of chemicals, pp. 375–381. In J. W. Pratt, Ed. Statistical and Mathematical Aspects of Pollution Problems. Marcel Dekker, New York.

Hoel, D. G., N. L. Kaplan, and M. W. Anderson. 1983. Implication of nonlinear kinetics on risk estimation in carcinogenesis. Science 219:1032–1037.

Jusko, W. J. 1972. Pharmacodynamic principles in chemical teratology: Dose-effect relationships. J. Pharmacol. Exp. Ther. 183:469–480.

Jusko, W. J. 1973. Pharmacokinetic principles in chemical teratology, pp. 9–19. In W. A. M. Duncan, Ed. Toxicology: A Review and Prospect. Proceedings, European Society for the Study of Drug Toxicity. Vol. 14. American Elsevier, New York.

Krewski, D., and J. Van Ryzin. 1981. Dose response models for quantal response toxicity data, pp. 201–231. In M. Csörgö, D. A. Dawson, J. N. K. Rao, and A. K. Md. E. Saleh, Eds. Statistics and Related Topics. North-Holland, New York.

Kupper, L. L., and J. K. Haseman. 1978. The use of a correlated binomial model for the analysis of certain toxicological experiments. Biometrics 34:69–76.

Lehman, E. L. 1975. Nonparametrics: Statistical Methods Based on Ranks. Holden Day, San Francisco. (457 pp.)

Mantel, N. 1969. Some statistical viewpoints in the study of carcinogenesis. Prog. Exp. Tumor Res. 11:431–443.

Munera, C. 1986. An empirical Bayes approach to risk assessment in certain reproduction studies. Ph.D. dissertation, Columbia University.

Rai, K., and J. Van Ryzin. 1984. A dose-response mode incorporating Michaelis-Menten kinetics, pp. 59–64. In American Statistical Association, Biopharmaceutical Section. Proceedings. American Statistical Association, Washington, D.C.

Rai, K., and J. Van Ryzin. 1985. A dose-response model for teratological experiments involving quantal responses. Biometrics 41:1–9.

Van Ryzin, J. 1975. Estimating the mean of a random binomial parameter with trial size random. Sankhya Ser. B. 37:10–27.

Van Ryzin, J. 1985a. Consequences of nonlinear kinetic dose-response models on carcinogenic risk assessment, pp. 119–132. In D. G. Hoel, R. A. Merrill, and F. P. Perera, Eds. Risk Quantitation and Regulatory Policy. Banbury Report 19. Cold Spring Harbor Laboratory, Cold Spring Harbor, N.Y.

Van Ryzin, J. 1985b. Risk assessment for fetal toxicity. Toxicol. Ind. Health 1:299–310.

Williams, D. A. 1975. The analysis of binary responses from toxicological experiments involving reproduction and teratogenicity. Biometric 31:949–952.

G

Empirical Modeling of the Toxicity of Mixtures

This appendix reviews some statistical methods for empirical modeling of toxicity at experimental doses. Suppose the goal is to study the toxicity of various mixtures of m chemical components. A major difficulty is the appropriate measurement of a toxic end point. Chemicals act on various organs and systems, so a proper measure must integrate several response factors. Assuming that an appropriate measure has been selected, the toxicity of a chemical is defined as the expected excess response in a population of organisms exposed to the chemical. If response is recorded on a binary scale, such as 0–1 or alive-dead, then toxicity is defined as the difference in the probability of, say, death between the presence and the absence of exposure to the chemical:

$$\text{tox} = P(\text{death with chemical}) - P(\text{death without chemical}).$$

Other response events could be substituted for death.

A mixture containing m chemicals will be represented by a vector $\mathbf{z} = (\mathbf{z}_1, \mathbf{z}_2, \ldots, \mathbf{z}_m)$, meaning that the mixture contains $\mathbf{z}_1$ dose units of chemical 1, $\mathbf{z}_2$ dose units of chemical 2, and so on. The goal of empirical modeling is to obtain a parsimonious description of the toxicity surface, $\text{tox}(\mathbf{z})$, over the experimental range. An empirical model will summarize important patterns in the data; this will allow some inferences to be drawn about the toxicity of the mixture and suggest avenues for further experimentation.

If the available data consist of direct joint measurements on the toxicity surface, then standard regression-analysis procedures can be used to model toxicity. In this setting, there is a broad class of powerful statistical tools accompanied by excellent computation software for exploratory investigation of the dose-response surface, for model fitting and diagnostic checking, and for experimental design. For toxicity data in the above format, the job of empirical modeling is relatively easy. A good reference text is Mosteller and Tukey (1977).

Many toxicity measurements, however, are quantal—for example, the proportion of animals that die prematurely when exposed to a chemical might be recorded. Such data do not follow a standard regression format (i.e., the underlying assumptions for regression theory are not satisfied). A useful approach to such data is based on extensions of the standard linear model developed over the last 20–30 years. For quantal response data, there are logistic and probit regression methods, which, in turn, are special cases of the more extensive collection of generalized linear models (GLMs) to be found in the book by McCullagh and Nelder (1983). For survival data, there are the Cox proportional hazards model and its extensions; relevant references are Carter et al., (1983), Cox and Oakes (1984), and Kalbfleisch and Prentice (1980).

An important aspect of the GLM and Cox model methods is that they can be implemented easily, given an adequate understanding of the application and theory of standard linear models (McCullagh and Nelder, 1983; Carter et al., 1983). In particular, the common exploratory and confirmatory data-analysis techniques developed for the standard regression setup have counterparts in the new framework. There are analogues of graphic exploration, variable selection, analysis of variance, residuals, and diagnostic checking. An exception occurs in experimental design, because most of the standard optimal experimental design literature on the linear model focuses on construction of designs that minimize the variance of some regression parameters. In the standard linear model, variance expressions are independent of the value of the response surface, so the optimal design is relatively easy to define. That does not happen with quantal response data, and the optimal design here depends on the unknown characteristics of the toxicity surface. As a result, the optimal design can be difficult to specify. It could be argued that, because the optimality criteria used to define designs are often very narrow, efforts spent on developing an optimal design might not always be justified.

An appealing approach to experimental design is that of Carter et al. (1983). Their work is concerned with finding favorable doses for chemotherapy. The typical measurements are either quantal response or survival times. The techniques that they propose involve sequential designs taken from response-surface methods (see, for example, Box et al., 1978). In several compelling examples given by Carter et al. (1983), the approach is shown to work very well. In the light of that experience, there is much to be said for using standard fractional-factorial, screening, and other familiar designs to explore toxicity surfaces, at least unless it becomes apparent that the designs are grossly inefficient. It is clear that further research needs to be done.

The purpose of this section is to show how the GLM strategy works. A complete account of the method, with many more examples, can be found in McCullagh and Nelder's (1983) book. The illustration of the use of GLM below demonstrates how the method preserves many of the most important aspects of ordinary regression analysis. Nonparametric extensions of GLMs and their use in estimating mixture toxicity are discussed further later. In the final section, the so-called biomathematical models introduced by Ashford and Cobby (1974) and Hewlett and Plackett (1979) are discussed. These models are developed from general quasimechanistic principles. Their implementation and interpretation, however, are more complicated than the GLM framework. It should be emphasized that the purpose of either model system (generalized linear or biomathematical) is to approximate the relationship between toxicity and dose over the observed dose range; extrapolation on the basis of these models without additional scientific insight might be inappropriate. Where extrapolation is required, the empirical model will, at best, suggest relationships that a scientist can use to postulate more refined theories.

ANALYSIS OF AN INHALATION EXPERIMENT ON RATS

To illustrate the use of the GLM method, we present a simple case study.

In one of a series of inhalation experiments carried out by B. C. Levin et al. (in press) at the National Bureau of Standards, groups of rats were exposed to mixtures of CO and CO_2 at high doses. Some of the experiments are described in Chapter 3 and Appendix C. For each mixture, the number of animals that survived was recorded. Preliminary data are listed in Table G-1 and plotted in

TABLE G-1 Mortality of Rats Exposed to CO_2 and CO[a]

CO_2, ppm	CO, ppm	No. Rats Dead	No. Rats Exposed
34,300	3,000	1	5
37,500	3,600	5	5
36,600	3,400	3	5
47,500	2,400	0	4
43,400	3,200	3	6
44,800	2,600	3	6
49,700	2,700	1	4
51,100	2,700	3	5
103,600	1,600	0	4
118,800	1,500	0	4
104,700	2,400	0	4
100,300	2,900	4	6
137,600	2,900	1	3
146,900	2,900	0	4
147,600	3,100	1	5
142,400	3,800	6	6
55,600	3,000	1	4
73,000	2,200	0	5
74,300	2,800	2	5
52,500	2,500	4	4
177,100	3,200	0	4
173,200	3,600	1	4

[a] Data from inhalation experiment, Levin et al. (in press).

Figure G-1. Note that the plot reveals some wide gaps in the design. Further experimentation might be directed toward filling the gaps, if estimation of the entire concentration surface were the objective of the experiment.

A close examination of Figure G-1 suggests that higher concentrations of CO_2 seem to decrease the toxicity of CO; hence, the ratio of CO to CO_2 might be an important explanatory variable. After some trial and error, a log-logistic model was found to give a reasonable fit to the data:

$$\log\{p(\mathbf{z})/[1 - p(\mathbf{z})]\} = \beta_0 + \beta_1(\mathrm{CO}) + \beta_2(\mathrm{CO_2}) + \beta_3(\mathrm{CO/CO_2}),$$

where $p(\mathbf{z})$ represents the probability of death after exposure to mixture $\mathbf{z}$ and the β_i are estimated under the model. The goodness of fit of this kind of model is measured by what is known as the scaled deviance (i.e., the analogue of a residual sum of squares). Here, the scaled deviance is 24.17 on 18 degrees of

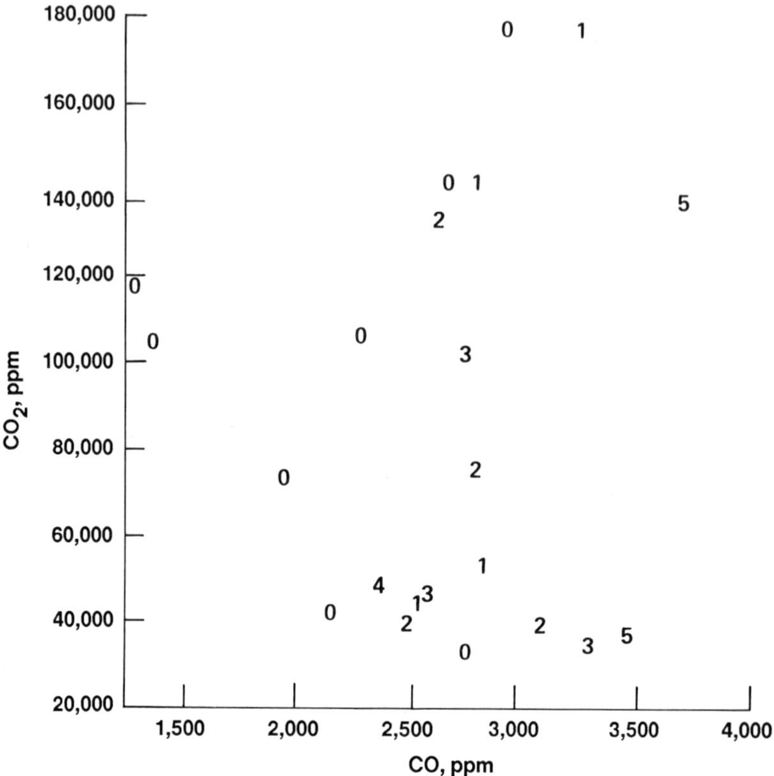

FIGURE G-1 Data from inhalation experiments (Levin et al., in press). Points are labeled 0–5, with 0 indicating that all animals survived at that concentration combination and other number indicating number of deaths.

freedom. For a standard regression model, the deviance reduces to the residual sum of squares. The scaled deviance has an approximate chi-squared distribution with a number of degrees of freedom equal to the sample size minus the number of parameters in the model. The deviance for our fit is larger than one would hope—the expected value for a chi-squared variate is equal to its number of degrees of freedom, so, because 24.17 is over 30% larger than the expected 18, one might worry about the adequacy of the model. The standardized residuals from the fit are plotted against the model-predicted response rates, *y* (the number expected to die at each observed combination of CO and CO_2), in Figure G-2.

If the model were correct, the residuals would approximate standard normal deviance. Note that the largest residual is more than two units away from zero. Given the sample size, that residual is somewhat suspect. The point corresponding to it is labeled "4" in Figure G-1. One can see that the point seems high relative to its neighbors. All the animals in that group were cannulated for blood withdrawal during exposure, and that might explain why the group is an outlier. Refitting the model with the outlier removed results in a model with a scaled deviance of 17.87 on 17 degrees of freedom. The reduction in the deviance of 6.3 units for 1 degree of freedom is dramatic. Residuals from the fit are acceptable, and the model provides a plausible summary of the data. The isoboles (curves of constant toxicity) for the fitted model are shown in Figure G-3.

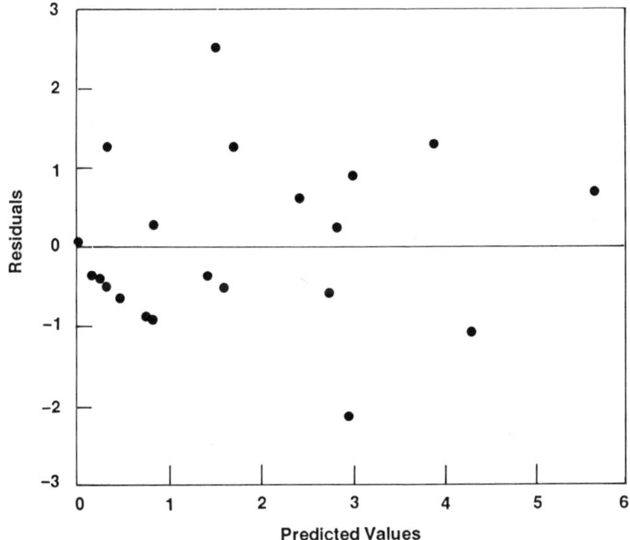

FIGURE G-2 Residuals from model fitted to all data. The largest residual corresponds to the point labeled "4" in Figure G-1.

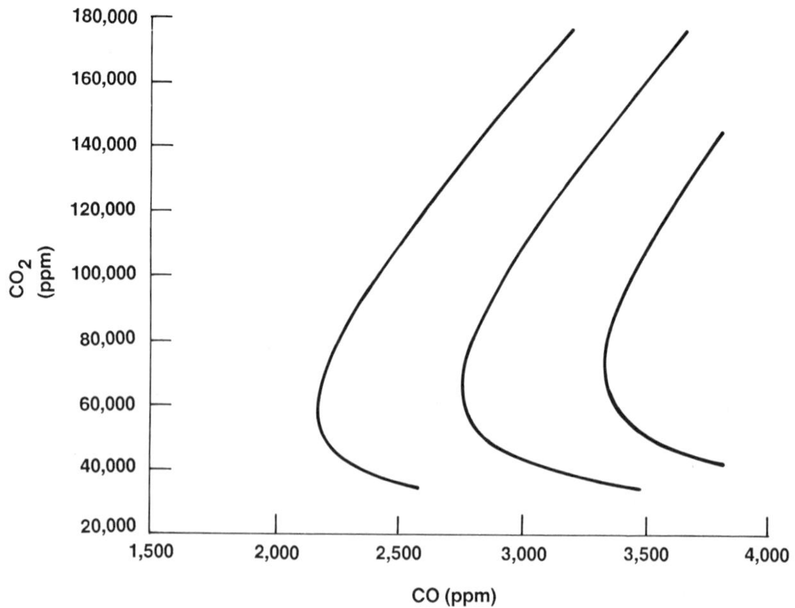

FIGURE G-3 Isoboles for fitted empirical model.

The 0.5 isobole corresponds to the set of combinations of CO and CO_2 that would lead to a 50% death rate in the population. The isoboles are highly nonlinear for lower concentrations of CO_2.

The example gives the essential flavor of the GLM approach to data analysis. Because the method preserves the common steps in the analysis of regression data, the analyst can learn to use it quickly. In the example, there were 22 data points and two explanatory variables; by modern standards, this is a modest size of data set. If the data set were smaller, it might have been more difficult to find an appropriate fit to the data; with more data, we would hope to get a more detailed description of the toxicity surface. The dangers of extrapolation are also shown by the example; clearly, animals cannot survive in a pure CO or pure CO_2 environment, as would be predicted by the model. However, for zero concentration of CO_2 combined with any concentration of CO (including concentrations above the LC_{50}, about 4,600 ppm), the model predicts zero mortality. Thus, extrapolation on the basis of this purely empirical model is inappropriate.

As is often the case in data analysis, progress would have been very slow without graphics. A variety of modern statistical software packages could have been used to implement our analysis. We used the GLIM statistical package that is distributed by the Royal Statistical Society and the S statistical package, which is an environment for data manipulation and graphics distributed by Bell Laboratories.

The analysis here used only functions of the concentrations of mixture components in estimating the mixture's toxicity. There is no reason, however, why other variables, such as mixture properties, could not be considered in the analysis.

A GENERALIZED CLASS OF NONINTERACTING MODELS

SIMPLE ADDITIVE MODEL

The simple model often used to describe the toxicity of a mixture is the additive model. The toxicity of the mixture is represented as

$$\text{tox}(\mathbf{z}) = \text{tox}_1(\mathbf{z}_1) + \text{tox}_2(\mathbf{z}_2) + \ldots + \text{tox}_m(\mathbf{z}_m),$$

where $\text{tox}(\mathbf{z})$ denotes the toxicity of the mixture with dose vector $\mathbf{z}$ and $\text{tox}_i(\mathbf{z}_i)$ describes the toxicity of the ith component of the mixture. If toxicities are measured as excess risks, the additive model says that the total excess risk is the sum of the excess risks associated with the components. This section shows that additive models constitute a large class of models, including many often used to address dose-response issues.

GENERALIZED ADDITIVE MODEL

An additive model might not be appropriate in the raw toxicity scale, but it might hold for some alternative scale.

$$g[\text{tox}(\mathbf{z})] = \phi_1(\mathbf{z}_1) + \phi_2(\mathbf{z}_2) + \ldots + \phi_m(\mathbf{z}_m),$$

where g is a transformation of the toxicity scale and $\phi_i(\mathbf{z}_i)$ describes the effect of the ith component on the overall toxicity of the mixture. If the responses are expressed as quantal data, then the transformation g may be a simple logit or probit, and the functions ϕ_i will typically be linear in dose or log dose. Some examples of generalized additive models follow.

In the additive-risk model, g is the identity function, $\text{tox}(\mathbf{z})$ is the excess risk associated with the mixture, and $\phi_i(\mathbf{z}_i)$ is the excess risk associated with the ith component. The public-health literature notions of additivity and synergism are based on this model.

In the multiplicative-risk model,

let $\text{tox}(\mathbf{z}) = 1 - Q(\mathbf{z})$ = probability of death, given dose $\mathbf{z}$ of the mixture, and

$Q_i(\mathbf{z}_i)$ = probability of survival, given dose $\mathbf{z}_i$ of component i.

The model assumes

$$Q(\mathbf{z}) = \prod_{i=1}^{m} Q_i(\mathbf{z}_i).$$

If $g(x) = \log(1 - x)$ and $\phi_i(\mathbf{z}_i) = \log[Q_i(\mathbf{z}_i)]$, the multiplicative model can be written as

$$g[\text{tox}(\mathbf{z})] = \sum_{i=1}^{m} \phi_i(\mathbf{z}_i),$$

which is in the form of a generalized additive model.

In standard probit/logit analysis, tox($\mathbf{z}$) is the probability of death, given dose $\mathbf{z}$ of the mixture; g is the probit or logit transformation; and the ϕ_is are linear in $\mathbf{z}_i$ or $\log(\mathbf{z}_i)$.

To show joint action with a multistage model where the exposures act at separate stages and only at one stage each, an example is taken from Appendix E. The model is

$$\text{probability of death, given dose } \mathbf{z} = 1 - \exp -[A(1 + B_1\mathbf{z}_1)(1 + B_2\mathbf{z}_2)]$$

where A, B_1, and B_2 are parameters. To fit that into the format of the generalized additive model, take g to be the complementary log-log transformation, $g(x) = \log[-\log(1 - x)]$, and let $\phi_i(\mathbf{z}_i) = \log(1 + B_i\mathbf{z}_i)$ and $\alpha = \log A$. Then

$$g[\text{tox}(\mathbf{z})] = \alpha + \phi_1(\mathbf{z}_1) + \phi_2(\mathbf{z}_2).$$

Additive models have served statisticians well for many years. In the case where the ϕ_is are parametric, the generalized additive models are a special case of the generalized linear models of McCullagh and Nelder (1983).

If a generalized additive model holds, the toxicity of a mixture can be assessed from knowledge of the effects of its individual components on the transformed toxicity scale.

Interaction depends crucially on the scale in which toxicity is measured. This dependence has led to debate as to whether interaction ought to measure departure from additive or multiplicative models, and so on. To rationalize the various interaction measures, it would help if the underlying models were always explicitly stated. In addition to removing some of the confusion, that approach would make it easier to plan experiments to detect interaction. In the context of generalized additive models, it seems reasonable that interaction ought to reflect the presence of first-order interaction (nonadditive) terms in the model. Hinkley (1984) has developed some analogues of Tukey's one-degree-of-freedom test that would aid in that purpose.

If functions g and ϕ_i have known parametric forms, such as g logit and ϕ linear, there is a well-established inference system for the model, because the

TABLE G-2 Influence of One Chemical on the Biologic Action of Another

	Site of Primary Action	
Interaction	Similar	Dissimilar
No	Simple	Independent
Yes	Complex	Dependent

model is in the category of GLMs. For these models, parameter estimation, confidence intervals, and procedures for analysis of diagnostic residuals are well established. Experimental designs for GLMs are only in an early stage of development (Steinberg and Hunter, 1984).

Given a large number of doses (say, more than 20), transformations g and ϕ_i can be nonparametrically estimated (see Breiman and Friedman, 1985, and Hastie and Tibshirani, 1986). More elaborate methods allowing for general interaction patterns between variables have been described by O'Sullivan et al. (1986). Although the reliability of the procedures might not be very high in small samples (fewer than 30 data points), the exploratory value of such methods can be very useful. In particular, parametric forms for g and ϕ_i are often suggested by those methods.

Quasibiologic Models

Hewlett and Plackett (1979) have described a more complicated system of empirical models (see also Hoel, 1983). Those models have been developed by Ashford and Cobby (1974) and Hewlett and Plackett (1959). The latter models, although widely used, make assumptions about the underlying biologic factors that might not always be appropriate; moreover, it is still awkward to apply them to any mixture more complex than binary.

The Hewlett and Plackett approach considers mixtures of two chemicals A and B whose action occurs at either of two sites S_1 and S_2. The joint action of the two chemicals is classified as similar or dissimilar, depending on whether the sites of action of the chemicals are the same or different. The chemicals interact if the presence of one affects the amount of the other that reaches its site of action. In the Hewlett and Plackett system, the joint action of the chemicals is classified as in Table G-2.

In a Hewlett-Plackett model, a toxic event is said to occur if the administered dose exceeds some tolerance or threshold in the organism and a probability distribution for tolerances in the population is specified. The Hewlett and Plackett (1979) formulation is developed in some detail for noninteracting chemicals. The general model is as follows:

$Q(\mathbf{z})$ = probability of survival, given dose $\mathbf{z}$ of the mixture.

The administered dose $\mathbf{z} = (\mathbf{z}_1, \mathbf{z}_2)$ gets translated to an effective dose $w = (w_1, w_2)$. The tolerance of the organism ω-scale is denoted $\omega = (\omega_1, \omega_2)$. A toxic event occurs (the organism fails) if the combined effect at either site of action exceeds some threshold. Formally, that is described as: For $0 \leq \nu \leq 1$, either or both of the following will hold:

$$w_1/\omega_2 + \nu w_2/\omega_2 > 1; \tag{G-1}$$

$$\nu w_1/\omega_1 + w_2/\omega_2 > 1. \tag{G-2}$$

Expression G-1 describes the combined effect of the mixture at site 1, and Expression G-2 the combined effect at site 2. Parameter ν measures the degree of similarity in the joint action of the chemicals. If $\nu = 1$, the action is a simple similar action; if $\nu = 0$, it is independent. For mathematical simplicity, an alternative form of the model is more often used. Here, a toxic event occurs if

$$(w_1/\omega_1)^{1/\epsilon} + (w_2/\omega_2)^{1/\epsilon} > 1, \tag{G-3}$$

where $0 < \epsilon \leq 1$. Parameter ϵ is a measure of action similarity. If $\epsilon = 1$, the action is a simple similar action; if $\epsilon = 0$, it is independent.

Given a specified distribution for l and a known relationship between w and $\mathbf{z}$, $Q(\mathbf{z})$ can be computed and, in principle, any unknown parameters can be estimated by the method of maximum likelihood. It is not known whether there is a software package that will conveniently implement the analysis. Model discrimination is difficult in the Hewlett-Plackett framework. Alternative specifications for the population tolerance distributions and the type of interaction can lead to indistinguishable forms for the toxicity surface (Hewlett and Plackett, 1979, pp. 49–50). Extension of the Hewlett and Plackett models to mixtures of more than two components has not been worked out in any detail. For consideration of mixtures with many components, such models would probably be very complicated.

Hewlett and Plackett models are based on receptor theory. (Hewlett and Plackett use "sites" in much the same way as Ashford and Cobby use "receptors.") Chemicals are assumed to "tie up" receptors at the site of action, thereby diminishing the activity at the site. For the case of a mixture of m chemicals, the activity at the site of action that results from a dose $\mathbf{z} = (\mathbf{z}_1, \mathbf{z}_2, \ldots, \mathbf{z}_m)$ is expressed as

$$ACT(\mathbf{z}) = \frac{1 + \sum_{i=1}^{m} \alpha_i a_i \mathbf{z}_i}{1 + \sum_{i=1}^{m} a_i \mathbf{z}_i}, \tag{G-4}$$

where a_i are association constants, $a_i \mathbf{z}_i$ corresponds to the effective dose of the ith component at the site of action, and α_i measures the effect on the activity due to the ith component of the mixture. The activity at the site of action is monotonically related to toxicity of the chemical:

$$\text{tox}(\mathbf{z}) = \text{f}[ACT(\mathbf{z})], \tag{G-5}$$

where f is a monotonic function. Thus, the greater the activity of the mixture at the site of action, the greater the risk of a toxic event.

The biologic sophistication in both biomathematical modeling systems is relatively modest, so one would be hard pressed to choose one system over the other solely on the basis of biologic plausibility. In addition, the parameters in both models are vaguely defined. In the receptor-theory models, if f were of known parametric form, a system of inference for the models could be obtained (see McCullagh and Nelder, 1983) by expressing a model in the format of a GLM [with a nonlinear predictor related to the activity function, $ACT(\mathbf{z})$]. For that reason, it could be argued that receptor models have a practical advantage over the models of Hewlett and Plackett.

SUMMARY

Conventional toxicity data, typically represented by dose-response functions, offer possibilities for modeling and analysis by many different approaches. A mixture by itself, administered as a unitary treatment, poses no novel problems in principle, and expressions of the function can be derived with standard methods. The standard method, however, is less versatile than the class of GLM for modeling mixture data. The models are congruent with, and extensions of, the more familiar methods of regression analysis and provide other desirable features. For example, normality and constant variance are not required for estimates of the random-error component, and interactions of systematic effects can be specified to hold on an additive scale by appropriate transformations of the scale. Also, GLMs are accessible through an extensive collection of computer programs that permit various approaches to be explored and that can model joint excess risks without reliance on arbitrary biologic mechanisms. Those methods, however, require larger data sets than are ordinarily collected by toxicologists; the precision of dose-response surface estimates improves as the number of data points increases.

Several quasibiologic models (e.g., Hewlett-Plackett and Ashford-Cobby) based on biologic mechanisms have been advanced and, in essence, are based on assumptions about the overlap and nature of interactions at common sites. Although some might be translated into GLM terms, their suppositions limit the ease with which they can be expanded into models for multicomponent mixtures.

REFERENCES

Ashford, J. R., and J. M. Cobby. 1974. A system of models for the action of drugs applied singly or jointly to biological organisms. Biometrics 30:11–31.

Box, G. E. P., W. G. Hunter, and J. S. Hunter. 1978. Statistics for Experimenters. John Wiley & Sons, New York.

Breiman, L., and J. H. Friedman. 1985. Estimating optimal transformations for multiple regression and correlation. J. Am. Stat. Assoc. 80:580–619.

Carter, W. H., G. L. Wampler, and D. M. Stablein. 1983. Experimental design, pp. 108–129. In Regression Analysis of Survival Data in Cancer Chemotherapy. Marcel Dekker, New York.

Cox, D. R., and D. Oakes. 1984. Analysis of Survival Data. Chapman and Hall, New York. (201 pp.)

Hastie, T. J., and R. J. Tibshirani. 1986. Generalized additive models. Stat. Sci. 1:297–318.

Hewlett, P. S., and R. L. Plackett. 1959. A unified theory for quantal responses to mixtures of drugs: Non-interactive action. Biometrics 15:591–610.

Hewlett, P. S., and R. L. Plackett. 1979. An Introduction to the Interpretation of Quantal Responses in Biology. University Park Press, Baltimore. (82 pp.)

Hinkley, D. V. 1984. Diagnostics for Transformable Non-Additivity. Technical Report No. 6. Center for Statistical Sciences, University of Texas, Austin, Texas.

Hoel, D. G. 1983. Statistical Aspects of Chemical Mixtures. Working Paper for SGOMSEC 3 Workshop, Guilford, England, August 15–19, 1983. National Institute of Environmental Health Sciences, National Institutes of Health, Research Triangle Park, N.C. [17 pp.]

Kalbfleisch, J. D., and R. L. Prentice. 1980. The Statistical Analysis of Failure Time Data. John Wiley & Sons, New York. (321 pp.)

Levin, B. C., M. Paabo, J. L. Gurman, and S. E. Harris. (in press) Effects of exposure to single or multiple combinations of the predominant toxic gases and low oxygen atmospheres produced in fires. Fundam. Appl. Toxicol.

McCullagh, P., and J. A. Nelder. 1983. Generalized Linear Models. Chapman and Hall, New York. (261 pp.)

Mosteller, F., and J. W. Tukey. 1977. Data Analysis and Regression: A Second Course in Statistics. Addison-Wesley, Reading, Mass. (588 pp.)

O'Sullivan, F., B. S. Yandell, and W. J. Raynor, Jr. 1986. Automatic smoothing of regression functions in generalized linear models. J. Am. Stat. Assoc. 81:96–103.

Steinberg, D. M., and W. G. Hunter. 1984. Experimental design: Review and comment. Technometrics 26:71–97.

Index

D

E

F

G

H

S

T

U

V

W